WITHDRAWN
UTSA LIBRARIES

WATER

A Study of its Properties, its Constitution, its Circulation on the Earth and its Utilization by Man

PLATE I.—THE HOOVER (BOULDER) DAM AND LAKE MEAD RESERVOIR are the highest dam and largest man-made lake in the world. They harness the Colorado River and make the lower part of the Grand Cañon accessible by boat.

Frontispiece.

WATER

A STUDY OF ITS PROPERTIES, ITS CONSTITUTION, ITS CIRCULATION ON THE EARTH, AND ITS UTILIZATION BY MAN

BY

SIR CYRIL S. FOX, D.Sc. F.G.S.

Lately Director, Geological Survey of India; Past President, Mining, Geological, and Metallurgical Institute of India, and Author of Engineering Geology, The Geology of Water Supply, *and other works on economic and engineering geology.*

GREENWOOD PRESS, PUBLISHERS
WESTPORT, CONNECTICUT

The Library of Congress has catalogued this publication as follows:

Library of Congress Cataloging in Publication Data

Fox, Sir Cyril Sankey, 1886-
   Water; a study of its properties, its constitution, its circulation on the earth, and its utilization by man.

    Reprint of the 1951 ed.
    1. Hydrology. I. Title.
[GB661.F65 1972]   551.4'8   75-138233
ISBN 0-8371-5590-8

Originally published in 1951
by The Technical Press, Ltd., London

Reprinted with the permission
of Philosophical Library, Inc.

First Greenwood Reprinting 1972

Library of Congress Catalogue Card Number 75-138233

ISBN 0-8371-5590-8

Printed in the United States of America

PREFACE

This treatise on water has been prepared as the intial volume of a series of books—The Geology of Water Supply, Water Treatment, Purification and Disposal of Sewage, Water Supply of Towns and the Construction of Waterworks, Percussion Drilling Rigs and Tools for Shallow Water Wells, Wells and Bore Holes for Water Supply, etc.—which the Technical Press, Ltd., have in preparation or have already published to meet the current need for information on the various aspects of water supply. In this volume the intention is to provide the reader with an outline of the science of water, both in the academic and practical aspects of the subject, and to include details regarding its characteristics, its occurrence, and its utilization. An effort has been made to deal with the subject as simply as possible without reducing the extent of the field covered, but avoiding elaborate scientific details.

Wherever possible, a practical example is given in the description, or a photograph is utilized to illustrate the matter under consideration. These pictorial examples are taken as far as possible from the English countryside or around the coasts and seas of the British Isles ; but most of the storage dams which are reproduced are from the great schemes of land-reclamation in the United States of America. The definite objective of this treatise is to draw attention to the immense importance of the control and use of water by first understanding the practical aspects of the material itself. In countries such as India, which are subject to alternate periods of dry and wet weather, it is a common experience for the land to be flooded by rivers during the wet monsoon and subsequently to be parched owing to lack of rain during the dry monsoon period, particularly in the hot months.

In such a country schemes of storage, either by impounding or deflecting underground (into water-bearing strata) are of an elementary type. In desert regions, such as the Sudan and Upper Egypt, the impounded water must rise above the level of the country to be irrigated so as to " command " a gravity supply. In the case of the Colorado River drainage, where the stream bed lies in a deeply cut gorge and the adjacent lands are arid regions the problem was complicated by the great floods which swept down the gorges at certain seasons and which constituted a menace to any feeble attempt at irrigation without proper control of the river. Any project in the Colorado River was of necessity on a stupendous scale, and so in the case of the building of the Hoover Dam and the impounding of Lake Mead the Bureau of Reclamation constructed the highest dam in the world and impounded the largest man-made lake. Lake Mead has a capacity of 32,359,000 acre-feet with a flood control reserve of 9,500,000 acre-feet. It is 115 miles long and has a water-spread of 146,500 acres and a maximum depth of 589 feet at the dam. The power plant capacity at the Hoover (Boulder) Dam is 1,835,000 horsepower and the spillways are able to discharge 200,000 cubic feet of water per second into the Black Canyon below the dam (or back into the river downstream).

The United States Department of the Interior, through the Bureau of Reclamation, also controls the construction of the dams and reservoirs of other projects, such as those of the Central Valley of California (with the Shasta Dam across the Sacramento valley and the Friant Dam astride the San Joaquin river); the Columbia Basin (with the Grand Coulee Dam 90 miles west of Spokane, in Washington, and the Bonneville Dam above Portland, Oregon); the Salt River (with the Roosevelt and other dams) in Arizona; the Rio Grande (Elephant Butte dam and reservoir), in New Mexico, and other great schemes to supply irrigation water, to improve navigational facilities, to provide for flood control, generate electric power, supply towns and industrial centres, safeguard fish and wild life, and to include means for recreation. The Tennessee Valley Authority, with its multiple purpose schemes on the Tennessee River from its confluence with the Ohio, near Paduacah, in Kentucky, and across western Tennessee into Alabama, and so up to eastern Tennessee and by its tributaries in North Carolina, by successive dams and locks, had established a pattern for all hydro-electric and other water-supply engineers to follow. The Tennessee River is, however, a different proposition from that of the Colorado River.

Here in England there is a tendency to bring these public service enterprises under central or national control, and this is the trend in the supply of water to towns and industrial centres throughout the British Isles. Although plans were made long ago for bringing water from Wales by pipelines to London, the metropolis still draws its water from the Thames through the great reservoirs near Staines and thereabouts. Birmingham has for many years obtained its supplies of good water from the Rhyaeder area of mid Wales. Liverpool has similarly drawn its supplies from Lake Vyrnwy (and Birkenhead from Alwen Lake), in Denbighshire (N. Wales), for several years. Manchester obtains its water from the Cumberland lakes—Thirlmere and Hawes Water. Derby, Leicester, Sheffield, and Nottingham, as a group, have a new supply from Derwent Valley Water Works. In all these cases the arrangements are in the hands of local bodies—the Metropolitan Water Board for London, Liverpool Corporation, Joint Water Board for Sheffield-Nottingham-Derby-Leicester, etc. In the case of the Derwent Valley scheme the consent of the House of Commons was obtained, and there is little doubt that all the major questions of water supply will be dealt with by the Ministry of Health and Lands, who, presumably, will take over all water rights.

There is no doubt that Government control may be of most benefit in a settled country by instituting a central authority for co-ordinating the various local water boards, and it is difficult to imagine the undertaking of vast irrigation or hydro-electric projects without Government assistance, but in unsettled regions or areas of small farms and ranches, particularly in Africa, Asia, and in Australia, there are continual problems of water supply which require a knowledge of water. The series of books which are in process of publication by the Technical Press, Ltd. (Gloucester Road, Kingston Hill, Surrey, England), should provide all the data and other information which a farmer or an engineer may require. It is hoped that this treatise on water may be helpful and of interest to a wide circle of readers, since it covers so wide a field of the characteristics, mode of occurrence, engineering experience, and modern projects for the utilization of water. It has been prepared for the practical engineering student rather than as a book of reference. More attention has been devoted to matters relating to large supplies of water than to details of a highly scientific nature, such as the origin of the atmosphere and the oceans, or the role played by super-heated waters deep within the earth's crust.

CYRIL S. FOX.

TUDOR HOUSE,
  QUEENSMERE ROAD,
    WIMBLEDON, S.W. 19.

TABLE OF CONTENTS

	PAGE
PREFACE	v
FOREWORD	xiii
INTRODUCTION	xvii

PART 1.—THE NATURAL HISTORY OF WATER

CHAPTER

I. THE CONSTITUTION OF WATER 3

Introductory Remarks	3
Forms of Water	4
Mode of Occurrence	5
Physical Properties of Water	7
Chemical Characteristics of Water	11
Heavy Water in Atomic Energy Development	16

II. THE DISTRIBUTION OF WATER 20

The Atmosphere	20
The Oceans and Seas	24
Lakes and Rivers	30
Polar Ice and Glaciers	34
Water in the Soil	35
Water Held in the Rocks	36

III. THE CIRCULATION OF WATER 40

The Heat from the Sun	40
Evaporation from the Sea and Land	41
Precipitation as Rain and Snow	46
Run-off Rainfall	51
Percolation and Infiltration	57
Weathering and Hydration	62
Hot Springs and Volcanoes	63
Summary of Rainfall Disposal	64

PART 2.—THE WORK DONE BY WATER

IV. EROSION OF THE LAND SURFACE 69

Denudation by Rain and Rivers	69
Waterfalls and Gorges	73
Action of Frost and Ice	76
Sediment carried by Rivers	78
Solids carried in Solution	79
Coastal Erosion by Currents	81
Erosion caused by Man	84

TABLE OF CONTENTS

CHAPTER | PAGE
V. THE ACTION OF UNDERGROUND WATER 85

    Percolation and Absorption 85
    Weathering of Rocks 88
    Underground Rivers and Springs 92
    Thermal Springs and Volcanoes 95
    Mineral Veins, Pegmatites, and Metamorphic Rocks . . . 100
    Radioactive Waters 102

VI. THE DEPOSITION OF SEDIMENTS 104

    Introductory 104
    The Stratified Sedimentary Deposits 104
    Fluviatile and Marine Deposition 108
    Shoals and Silting 110
    Deposits of Saline Residues 114
    Deposits from Hot Springs and Fumaroles 117
    Tinstone and Fossil Wood 119

PART 3.—THE UTILIZATION OF WATER

VII. GENERAL CONSIDERATIONS 123

    Hydrographic Considerations 123
    Collection of Data 124
    Water Supply Factors 125
    Danger from Contamination 126
    Stream Gauging and Storage 128
    Measurement of Silt 129
    Wells and Springs 130
    Artesian Water 130

VIII. WATER SUPPLY ENGINEERING 132

    Multiple Purpose Schemes 132
    Power from Rivers 135
    Power from Natural Steam 136
    Volcanic Energy 138

IX. CONCLUDING REMARKS 140

    The Radioactive Water in Bikini Atoll 140
    Water Rights 140
    The National Aspect of Water Supply 142

INDEX 145

ERRATA

p. 27, l. 7. *For* (p. 29) *read* (p. 31).
p. 30, l. 8. *For* p. 23 *read* p. 25.
p. 55, l. 31. *For* page 31 *read* page 33.

PHOTOGRAPH ILLUSTRATIONS

PLATE		FACING PAGE
I.	*Frontispiece.*—The Hoover (Boulder) and Lake Mead (the impounded reservoir).	
II.	Ground Mist—water vapour, England	16
III.	A River beginning at a Glacier.	16
IV.	An Iceberg melting in the North Atlantic	17
V.	The Berezov Mammoth from Siberia	17
VI.	A Radioactive Hot Spring in Abyssinia	32
VII.	Tufa Deposits of Hot Springs in Wyoming	32
VIII.	Old Faithful Geyser, in Yellowstone Park	33
IX.	The Valley of Ten Thousand Smokes, Alaska	48
X.	Natural Bridge (Grosvenor Arch), Utah	49
XI.	Grand Canyon, Colorado River, Arizona	64
XII.	Niagara Falls of America and Canada	65
XIII.	Victoria Falls, on the Zambesi, Rhodesia	65
XIV.	Chasm of the Victoria Falls, Rhodesia	80
XV.	Coast Erosion on the Norfolk Shore	81
XVI.	Heavy Seas pounding Hastings promenade	81
XVII.	Flooded Valley of the Severn River, England	96
XVIII.	Flood-covered falls at Wenatchee, Columbia River	96
XIX.	Grand Coulee Dam, Columbia River, Washington State	97
XX.	Bonneville Dam, Columbia River, Oregon State	112
XXI.	Norris Dam, Clinch River, Tennessee (T.V.A.)	113
XXII.	Wheeler Dam, Tennessee River, Alabama (T.V.A.)	113
XXIII.	Parker Dam, Colorado River (150 miles below the Boulder Dam)	128
XXIV.	Imperial Dam De-silting Works, Colorado River (below the Parker Dam)	129
XXV.	Steam Escaping from Vesuvius, Italy	129

TEXT-FIGURES

TEXT-FIGURES		PAGE

1. Geological Section, showing Coastal Problems of Water Supply 58
    - A. Fresh water under sand dune.
    - B. Water held up by Dolerite dyke.
    - C. A failure in Granite.
    - D. Success in striking joint planes.*

2. Birds-eye view of Victoria Falls and Gorge on the Zambesi River, Rhodesia 74
    (After H. B. Maufe, see p. 71.)

3. Geological Section of Silted-up Valley 93
    - A. Shallow well in alluvium on bank.
    - B. Deeper well after boring into bed (sand).
    - C. Deep boring into gravel of old bed.
    - D. Shaft to meet stratum which taps the old bed and up which a drivage will go.*

4. Profile and Sketch Plan of Tennessee Valley . . . 133
    (After diagram prepared by the Tennessee Valley Authority to illustrate the T.V.A. multi-purpose project now in operation in the Tennessee Valley. See p. 126.)

* For details relating to " finding water " and related questions of geological factors the reader is referred to the author's book *Geology of Water Supply* (1949).

FOREWORD

As has already been explained in the Preface, this treatise on water is to be regarded as the first of a series of books on the practical aspects of water supply. It is written in as non-technical a manner as possible, to be easily read by all, but it is technically correct and up-to-date. Besides dealing with the physical and chemical properties of water as it normally occurs as rain and snow, or obtained from lakes and rivers, its emergence from deep within the earth as hot springs and volcanic eruptions, is outlined. The possibility of some original or " virginal " water being given up from the magma within the earth's crust is considered, but the origin of the earth's water is regarded as too intimately related to the origin of the earth itself to be treated fully in a treatise on water.. It is a problem of modern physical chemistry, as is " heavy water ", which is briefly dealt with under the subject of using water as a raw material for the derivation of atomic energy by nuclear fission.

Next in importance to the oxygen of the air water is the substance without which all life on the earth would perish. Yet it is nothing short of marvellous that a substance of such abundance and so simple in constitution should possess the many characteristics which guide the processes governing the changes on the face of the earth. By its presence in the atmosphere it tempers the sun's heat ; the rain that falls scours the hills and carries the sediment into the river valleys and deltas ; the streams remove almost unbelievable amounts of solid matter from the rocks, in solution, and carry these dissolved materials into the sea to concentrate them there. The water that percolates into the rock-crust of the earth may penetrate to great depths and take part in the formation of mineral deposits, and emerge as thermal springs. In its deeper reactions it may lower the melting points of highly heated rocks and be discharged as steam during volcanic eruptions, such as the remarkable explosion of Krakatoa.

In the Polar regions water accumulates as great ice caps which may influence climatic and geographical changes by their increase or decrease in extent and thickness. It is known that the Pleistocene Ice Age, which was at its maximum spread 250,000 years ago in the northern hemisphere, loaded Europe and North America with immense ice sheets. As these have withdrawn and the continents have become unloaded, the land has not only been freed of ice, but has suffered a rise of level, an amelioration of climate, and a change in its river drainage system. Geological investigations have shown that there have been earlier Ice Ages in the long history of the earth, also accompanied by changes of climate and in the distribution of land and sea. The Ice Age of the southern hemisphere, Gondwanaland, occurred 200,000,000 years ago, and, like that of the relatively recent Pleistocene Ice Age, was of relatively short duration.

The presence of water on the earth may be traced back 1,500,000,000 years or so, when the earliest sedimentary deposits were laid down in the waters of those Azoic times. It is not until about 500,000,000 years ago that life appears to have been relatively abundant in the earliest Palæozoic (Cambrian) seas. Unmistakable evidence of the existence of man-like animals (*Pithecanthropus erectus* from Java, our predecessors, not our ancestors) is not found until the Quaternary period and probably only 500,000 years ago. No living member of the anthropoid apes is the ancestor of man, but the human race (*homo sapiens*) has descended from some unknown member of the group, probably subsequent to the Java specimen mentioned above. True man was evidently a contemporary of the woolly mammoth (*Mammuthus* or *Elephas primigenius*) at the close of the Pleistocene Glacial epoch in Northern Europe perhaps less than 50,000 years ago.

A complete carcass of a mammoth (see photograph) was found in 1900 on the left bank of a small stream (Berezovka), a tributary of the Kolyma river, 60 miles within the Arctic Circle and 200 miles northeast of Sredne-Kolymsk, Yakutsk Province, Siberia. It is now set up in the Zoological Museum in Leningrad. These animals have been extinct for several hundred years, and it would thus be reasonable to suppose that the Berezov mammoth has been preserved intact (its flesh was fed to the sledge dogs) for a very long time. The value of ice for meat packing and preservation is thus demonstrated in a very convincing manner. Other operations performed by water over a long period are those of erosion, such as the carving out of the Grand Canyon by the Colorado River, in Arizona (see Photograph XI), the Niagara Falls (Photograph XII), the Victoria Falls on the Zambesi River (Photographs XIII and XIV), etc.

The astonishing power exerted by a flood of rushing water, both in scouring and in transporting material, is rarely fully appreciated even to-day. The examples of the eroding action of water over a long period of time, mentioned above, give no idea of the ability of a stream in flood in a narrow gorge like the Black Canyon, in which the Hoover or Boulder Dam was founded. While founding the dam it was discovered that during high floods the entire silt, to a depth of 40 feet and perhaps much more, must be in movement, and must be replaced as the current slackens and redeposits silt. Movement of silt to such depths is now suspected in the beds of many large rivers, and probably explains the undermining of piers and abutments in those cases where a flood has breached a bridge.

The control of flood waters is an engineering problem quite as important as the prevention of landslips. A large slip during a period of heavy rain may supply a flooded river with so much material that the subsequent deposition at the head of a reservoir may seriously reduce the capacity after a few hours' discharge. In the case of the All American Canal, which carries water for irrigation in the Imperial Valley in Southern California, the desilting works at the Imperial Dam are capable of handling 15,000 cubic feet of water per second. I should like to mention that Professor Orville L. Bandy, of the University of Southern California, procured and sent me a copy of the Report on " The Colorado River ", which gives a comprehensive account of the development of the water resources of the Colorado River basin for irrigation, power production, and other beneficial uses in Arizona, California, Colorado, Nevada, New Mexico, Utah, and Wyoming. I am very grateful to Professor Bandy, and also to Mr. Howard Gould, for the present of this valuable brochure which shows that a natural menace has indeed become a national resource.

Besides controlling the Colorado from the Hoover Dam and Lake Mead, and supplying water for the irrigation of the Imperial Valley from the All American Canal (from the Imperial Dam), Los Angeles, and many more cities are supplied with water from the Colorado River Aqueduct which takes off at the Parker Dam. In addition to these great works, and to the generation of electric power, the Bureau of Reclamation, United States Department of the Interior, have in hand the Salt River Project in Arizona, the Columbia Basin Scheme in Washington and Oregon, the Central Valley Project in California, the Rio Grande Scheme in New Mexico, and other projects. I am most grateful to Mr. K. K. Young, of the Engineering Section of the Bureau of Reclamation, for pamphlets and photographs of the above-mentioned projects. He has also been kind enough to have communications directed from the Bureau of Reclamation to the Tennessee Valley Authority, the Bonneville Power Administration, and the Corps of Engineers for additional data on my behalf.

As a result of the above-mentioned requests, I have been supplied with photographs and other information by Mr. Maurice Henle, Chief of the Information Section, Tennessee Valley Authority, Knoxville, Tennessee ; from Mr. Wilbur D. Staats, Chief, Information Service Section, Bonneville Power Administration, Portland, Oregon, U.S. Department of the Interior, and from Lieut.-Colonel M. L. Webster, Corps of Engineers, Department of the Army, Washington. I am very much obliged to all. I have also been placed under considerable obligation to the National Geographical Society, Washington, D.C., for a carefully chosen series of photographs of natural phenomena—geysers, gorge, stone arches, waterfalls, etc., from which I have made selections for the illustrations reproduced in this treatise (Photographs VIII–XI). I should mention that I have secured a number of the photographs through the kindness of Mr. William D. Clark, London Editorial Director, *Encyclopaedia Britannica*. To all these generous friends I have to add grateful thanks to the United States Information Service in London, from which I obtained upwards of thirty photographs.

The High Commissioner for Southern Rhodesia very kindly sent me photographs of the Victoria Falls on the Zambesi; Mr. D. Wilkinson, of *Water and Water Engineering*, placed at my disposal the indexes of his journals covering ten years since 1939 and offered me the loan of any journals I might need to consult. Mr. J. Foster Petre, of *Engineering*, also provided me with a list of references and offered me the loan of blocks for any illustrations I might choose to reproduce from those in his office. Mr. A. H. Saville, Distribution Manager, Kemsley Newspapers, Ltd., sent me a fine collection of " Mist and Cloud " photographs (see Photograph II). It is impossible to express adequately my appreciation of all this kindness. I am grateful for the active and prompt help I received from these important journals, and to them I have to add several others—the Associated Press, Ltd., 85 Fleet Street; the Central Press Photos, Ltd., 119 Fleet Street; the Sport and General Press Agency, Ltd., 4 Racquet Court, Fleet Street; The Graphic Photo Union, Gray's Inn Road, London—to mention only a few of those who searched their collections to provide me with pictures for my illustrations of " water ".

In making these acknowledgments it is a great pleasure to record the unfailing patience and constant encouragement I have received from the Technical Press, Ltd., the publishers, and I must also thank the printers, Stephen Austin and Sons, Ltd., for the clear reproduction of the illustrations and the excellence of their proofs.

INTRODUCTION

It will be seen from the table of contents of this treatise on water that an effort has been made to deal with the subject in the widest manner without entering into exhaustive details. It is hoped that this book may meet the need of civil engineers and others, for a modern book dealing with the scientific and practical aspects of the study and utilization of water. It has been difficult to find a title, since it covers the whole subject of water. The name " Hydrography " naturally comes to mind as most appropriate from a scientific point of view, but the title " Hydrography " has been, and still is, used in a restricted sense for that branch which deals with surveys for charting coasts and rivers and navigable waters, and for providing information on soundings, currents, and tides, etc., for navigational purposes.

Perhaps a more comprehensive title would be " Hydrology " for the science of water, its properties, laws, phenomena, etc. This name has been employed by the National Academy of Sciences, Washington, in their great compilation *Physics of the Earth*, vol. ix (1942), which is called " Hydrology ", but in a restricted sense also, since it does not include oceanic waters. This is dealt with in vol. v (1932) under the title " Oceanography ". They also, and quite rightly, have a further volume, iii (1931), on meteorology. In the volume entitled " Hydrology " the editor, Dr. O. E. Meinzer, mentions that the name " Hydrology " has been employed by geologists for studies of underground water supply and continues to be so used, instead of some such title as " Hydrogeology " or " Geohydrology ". However, these titles are not attractive. Among other names for the general subject of water there is " Hydrognosy " for the history and description of the waters of the earth. One could also coin a name such as " Hydrollography ", from *hydrol*, the name of the simple molecule, H_2O, of water, but the word is cumbersome.

In the circumstances it was of interest to look into the etymological aspect of the English word *water*, or at least to glance over its derivation or relationships. *Water* has the same meaning as the Old Teutonic *watar*, which in turn resembles the Russian word *voda* (written вода, *water*), from which the name *vodka* is evidently derived. Then comes the Greek word *hudor* (written ὑδωρ, *water*), which appears to be closely related to the Russian, and from which comes the prefix *hydro-* (written ὑδρο) for all terms referring to water. And lastly there is the Sanskrit word *udan* (written उह्म्रण, *water*), derived from the Sanskrit root *ud-*, having the meaning " to be moist ". There would thus appear to be reason for believing that the word *water* is of ancient lineage, and like the other words for water mentioned above, is of Indo-European origin and probably derived from the Sanskrit root *ud-*. It may be mentioned that the root word *ud-* can also be written *wd-*, *wod-*, and *wed-*, and have affixes such as *or*, *er*, *on*, and *en*. Therefore, seeing that the word *water* is simple and clear and derived from an Indo-European root, it seems most satisfactory to use this word as a title to a treatise dealing comprehensively with the subject of water.

For convenience of description the subject of water is dealt with in the following pages under three main headings :—

Part I.—*The Natural History of Water* : dealing with the physical properties, chemical constitution, distribution, and circulation of water on the earth.

Part 2.—*The Work Done by Water* : describing the geological aspects of erosion, solvent action and leaching, sedimentation and precipitation from water.

Part 3.—*The Utilization of Water* : discusses problems of collecting data, obtaining water supplies from rivers, wells, and springs, the subject of artesian water, engineering questions of storage reservoirs and power from falls, and various other aspects of the treatment of sewage, the purification of water, desilting, and multiple projects.

A fundamental consideration in the study of the earth's water resources is the so-called " Hydrolic Cycle " which refers to the circulation of water on the earth. The heat of the sun is the source of energy which, besides warming the land and ocean surfaces, evaporates water from these surfaces. The water vapour so formed is taken up by the air and carried by winds. When the moisture-laden winds lift, clouds are formed and where the resultant cooling is sufficient, as against high mountains, the vapour condenses and falls as rain or hail or snow. The precipitated water, rain, falls on the land and flows into streams and rivers and thus returns to the sea. And the process is repeated again and again. Judging by the age of the strata in the earth's crust these processes have operated for more than 1,500,000,000 years, or 1,000,000,000 years previous to any evidence of plant or animal life on the earth.

Physicists have computed that the earth receives only $\frac{1}{2000}$th millionth part of the energy radiated from the sun, but that even this minute fraction represents 1·95 calories per minute per square centimetre on its outer atmosphere. The heat that comes through the atmosphere to the land and sea surface, in the tropics, is estimated at 1·33 calories per minute per square centimetre. There is a loss of about 10 per cent by reflection back into the atmosphere. This means an available 1·20 calories per minute per square centimetre for heating the surface of the ocean and land on which the sun's rays fall. In terms of active energy this amount is theoretically equal to 4,500 horse-power or 3,356 kilowatts continuously generated from one acre exposed to the sun's heat on sunny days.

These processes of evaporation, transportation, and precipitation of water have, as already stated, been operating from the beginning of geological time (reckoned since the earliest sediments were deposited). The falling rain and scouring streams have carried solid matter, both in suspension and solution, from the hills into the valleys and to the deltas and the sea. Even if we allow a removal of a thickness of 1 inch of the land surface in 500 years, the total erosion in the past 1,500,000,000 years would be more than 47 miles (250,000 feet). If we restrict the time since the beginning of the Palæozoic (Cambrian), about 500,000,000 years ago, the total is a thickness of nearly 16 miles (roughly 84,000 feet). The Tertiary strata in Assam deposited over a period of perhaps 40,000,000 years are 40,000 feet thick, but are perhaps an exceptionally rapid deposition in a marine gulf. The Mississippi discharges 400,000,000 tons of sediment into the Gulf of Mexico yearly from a drainage area of 1,250,000 square miles (without taking into account 100 tons of dissolved matter per square mile from its catchment).

Such estimates are, of course, generalizations, since it is not possible to say if the rate of erosion has been at all similar during past ages. The rate varies in different parts of the world's surface to-day. However, it is clear that if the superficial area of the earth's surface is taken at 197,000,000 square miles (with a land surface equal to 57,000,000 square miles and an average altitude above sea level of 2,500 feet, and an oceanic area of 140,000,000 square miles with an average depth of 12,500 feet), and also that if the distribution of land and sea were the same 20,000,000 years ago, say at the close of Miocene times, as they are to-day, all the land would have been washed down to sea level. It is difficult to understand how so much land remains to-day if erosion has been as active as 1 inch of surface depth in 500 years, unless, of course, the sea has diminished in volume. In this connection, however, a claim is made that the Pacific Ocean level is rising at the rate of 1 inch in 50 years, while the Atlantic is rising at twice this rate. If these claims are true it must be that water is being given up from the sub-crustal region, and that this addition is " virginal water " which has been held in the rock magma since the beginning of the earth.

In making estimates of the water resources of the earth it is necessary to consider two accounts : one, the water that is more or less constant in the atmosphere, the oceans, and on the land ; and the other, the water that is in circulation (the hydrolic cycle). Among the chief items of the water in the " fixed deposit " are : (i) that in the atmosphere ; (ii) in the oceans or hydrosphere ; (iii) in the polar ice caps and glaciers ; (iv) in the lakes and rivers ; (v) in the soil and by plants and animals ; (vi) in the rocks and strata within a depth of 12,500 feet below sea level, and (vii) in the sub-crustal rocks between depths of $2\frac{1}{2}$ and $12\frac{1}{2}$ miles below mean sea-level. The " current account " of water in movement may be dealt with under the heads : (*a*) Evaporation from the Ocean, Lakes, and Land ; (*b*) Precipitation as Rain and Snow, etc. ; (*c*) Run-off Water from Rainfall, Melting Snow and Ice, which Flows to the Sea ; (*d*) Percolation and Absorption, and (*e*) Discharge from Springs, Geysers, and Volcanoes. The time interval between evaporation, rainfall, and run-off into the sea may be very short, while the interval between percolation into the rocks and emergence from springs may be appreciable, and the period from its being held by freshly deposited sediment to its discharge as steam from a volcanic eruption might be very considerable.

Some rough calculations of the " fixed account " of water are of interest and give the following approximations :—

(i) Moisture in the atmosphere, assuming 3 grains per cubic foot of air and presuming half the moisture as below an altitude of 7,000 feet, say $1\cdot33$ miles, works out to about 3,600 cubic miles or a layer $1\cdot16$ inches over the surface of the earth.

(ii) Water in the oceans and seas, reckoning an area of 140,000,000 square miles and a mean depth of $2\frac{1}{4}$ miles, about 315,000,000 cubic miles, which, if all the land was levelled along the sea floor, would cover the earth's surface to a depth of about 9,000 feet and approximately 600 feet above the existing mean sea-level.

(iii) Water in the ice caps and glaciers, taken as covering 8,000,000 square miles to a mean depth of 528 feet, is 800,000 cubic miles, which if spread uniformly over the earth's surface would cover it to a depth of about $21\frac{1}{2}$ feet (or raise the level of the existing oceans by at least 30 feet).

(iv) Water in lakes and rivers and marshes on the land, taken as equivalent to 500,000 square miles with a mean depth of $528\cdot0$ feet, gives 50,000 cubic miles and if spread over the land surface would cover it with a layer about $55\cdot68$ inches deep, increase the height of the oceans by $22\cdot65$ inches, or form a layer $16\cdot0$ inches deep over the earth's surface.

(v) It is impossible to make any reliable calculation for the water held in the soil and by plants and animals, but the amount is probably much less than that held by the atmosphere, say equivalent to a layer 1 inch deep on the land surface.

(vi) The water held by the rocks under the land to a depth of 2,500 feet (sea-level) computed with a pore space of about 4 per cent represents 1,084,000 cubic miles and would cover the land surface to a depth of 100 feet nearly. The water held by the rocks under the land from sea-level to a depth of 12,500 feet (say $2\frac{1}{4}$ miles), reckoned with a pore space of 1 per cent, amounts to 1,285,000 cubic miles and would cover the land surface to a depth of 125 feet approximately. The total water held in the rocks, as above, if yielded up and spread over the earth's surface, would cover it to a depth of nearly 65 feet.

(vii) The water held by the heated, sub-crustal rocks, probably in combination, reckoning at only 1 per cent, amounts to 19,700,000 cubic miles or a layer 10 miles deep from −12,500 feet below sea-level. This strongly held water, if yielded up, would cover the earth's surface to a depth of 528·0 feet. Some of this water may be original or " virginal ", but there can be little doubt that a considerable proportion must be meteoric water, trapped in the pore spaces of the sedimentary rocks which have become deeply buried and have retained this small proportion of the water in which they were deposited, either in the sea or river valleys.

Calculations for the " current account " of water in circulation are more difficult because average figures are unreliable. However, the following approximations may be of interest, but it is to be remembered that evaporation must vary greatly between the tropics and the frigid zone, even on the surface of the oceans, and markedly so in the case of enclosed seas at different latitudes. The estimates are :—

(*a*) Evaporation from the oceans and seas for an area of 90,000,000 square miles (out of 140,000,000) at an average of 30 inches annually, is about 42,600 cubic miles ; evaporation from ice and snowfields, for an area of 8,000,000 square miles at an average of 24 inches a year, is roughly 3,030 cubic miles ; the evaporation from 1,000,000 square miles of lakes and river surface at 36 inches a year amounts to nearly 570 cubic miles. The total thus amounts to 46,200 cubic miles of water (189,000,000,000,000 tons) evaporated by the sun annually, and equivalent to about 4 feet of rainfall on the land surface (57,000,000 square miles).

(*b*) Precipitation as rain and snow : the rainfall is taken as 36 inches a year on a land surface of 36,000,000 square miles, which amounts to 20,000 cubic miles annually. The snowfall is reckoned at 36 inches on 8,000,000 square miles a year, which is roughly 4,500 cubic feet (for land areas). The rainfall and snowfall on the ocean surfaces will probably make up the difference between 24,500 cubic miles and the total under (*a*), 46,200 cubic miles, which would be 21,700 cubic miles.

(*c*) The run-off water from the rainfall and snow on the land (allowing for re-evaporation, both of snow and rainwater) is very commonly taken as a third of the rainfall, and this proportion might be accepted as the quantity which travels to the sea and replenishes the lakes. Since the total rainfall and snowfall on the land is 24,500 cubic miles annually, the run-off flow is reckoned at about 8,200 cubic miles for the entire land surface of 57,000,000 square miles.

(*d*) The loss by percolation and absorption is also reckoned as one-third of the rainfall and snowfall on the land. This is approximately

8,200 cubic miles. It is presumed that the remainder of the rainfall (8,100 cubic miles) is partly re-evaporated, partly lost into the upper air, and partly held by plants and minerals in the soil.

(e) The discharge of springs, geysers, and volcanoes cannot be approximated with any degree of reliability since many great springs emerge in the sea bed; reliable estimates of the steam from geysers are not available, and the amount of vapour given up by volcanoes is quite unknown. There is no doubt that the quantities are large in each case, but their approximation will require a special investigation.

Geologists are agreed that at depths of about 10 to 12 miles the rocks are at such temperatures and under such pressures as to be in a plastic condition. With a temperature gradient of 1° C. rise for every 100 feet of descent, the temperature at a depth of 12 miles must exceed 650° C., which, with the pressure involved and the presence of combined water, is probably near the fusion point of most igneous rocks—granites as well as basalts. With this rubber-like stratum, ready to become molten with release of pressure or influx of water, the earth's crust is subject to bending under excess loading, say, the great thickness of ice which covered the northern hemisphere during Pleistocene times. With the withdrawal of the ice sheets and the consequent unloading the crust will rise. This has happened in North America in the region of the Great Lakes, and the slow uplift appears to be still in progress and affects the drainage of the region. The Ottawa river, for example, once drained the lakes, which now discharge into the St. Lawrence river. In the future, if the land continues to rise, the Mississippi will draw its waters from Lake Superior, while its tributaries, the Illinois and Ohio, respectively, will tap the other lakes, Michigan and Erie. Similarly, the erosion of mountain ranges, such as the Urals of Russia or the Appalachians of America, results in unloading and rise, while deposition, such as in the deltas of the Ganges and Mississippi, causes loading and continued sinking. Thus, water is perhaps the most powerful of all geological agents in effecting changes in geography.

Geographical changes affect the climate, and there is no doubt that water, as rain or snow, fogs and cloud, influences climate in a marked degree, notwithstanding the latitude of a country. Many suggestions have been made for controlling (improving) the weather, but the scale of operations is so large that it is still true that the best way to control the climate is to do so inside a house. Nor does there seem to be any practical solution of the problem of producing artificial rain from the sky in dry weather. One meteorological expert has made the statement: " There are two kinds of rain-making schemes—those that are too expensive to use generally and those that don't work." There is no doubt that the solid (frozen) carbonic acid, commonly known as " dry ice ", has been successfully used to precipitate rain where the weather conditions (humidity) were suitable for such experiments. However, if cheap electric power can be made available it would be more satisfactory to pipe water from an assured source to a deficient area. In 1937 a scheme was nearing completion in the Dnepropetrovsk Province of Soviet Russia whereby a motor-driven high pressure pump was to supply sprinklers from a pipe system. The spray from the sprinklers was to be wind-borne to the fields and so irrigation by artificial rain was designed. One such unit was reckoned capable of irrigating between 175 to 325 acres in a season, and the estimated cost of the installation was 120 roubles per acre per millimetre of rain, with an operating cost of from 12 to 28 kopeks. A rouble at that time was rated at about 10d. but was actually exchanged for 5d. The kopek would be 0·1d. Since the water was carried by pipes there would be no loss by leakage and evaporation, and over-irrigation by flooding was also avoided.

Although generalizations are permissible in reckoning the proportion of the rainfall in terms of run-off flow, evaporation, and percolation, etc., it is of considerable importance that these fractions should be determined where large-scale storage of water is contemplated. On the one hand, it may be found that the percolation from the floor of a reservoir is so great that the project is uneconomic and even a failure, while, on the other hand, if the run-off is larger than anticipated, there will be a constant condition of flooding and the spillway may not be able to deal with a high flood. The result may be a washaway of serious dimensions. For the reasons stated it is essential that great care be exercised in the collection of data for a water supply project—geological, meteorological, and engineering (such as stream gauging, etc.). In hot countries, subject to dry conditions and high winds, the loss by evaporation from a small reservoir may be as much as 60 to 72 inches a year. In hilly country where the strata are soft or unstable a landslip during a night of heavy rain may provide enough silt to the flooded rivers to cause very serious silting in a reservoir. Carelessness in founding a storage dam may easily lead to the thrusting away and failure of the dam with the discharge of a disastrous flood down the valley. Notwithstanding all precautions in the meteorological, geological, and engineering aspects of a great storage scheme, it is still possible for the canals and distributories below to be faulty and the leakage therefrom to cause waterlogging, and thus render the adjacent country unhealthy.

As industrial developments continue and the problems of agriculture become more urgent, in all countries attention is being focused on the subject of irrigation by the reclamation of arid lands. It is not possible to give, in general terms, any accurate idea of the amount of water that may be needed per acre, as there are so many factors—

climate, soil, and crop, also the method of irrigation to be adopted. The quantities may vary from 6,000 tons (224 gallons to one ton, or 36 cubic feet) in some cases to 3,000 tons (for basin types, such as that in Egypt) to 600 tons (at each watering in the perennial canal method such as that in India) to perhaps as little as 200 tons (in the Russian " artificial rain " system at each watering) per acre. However, it must be understood that notwithstanding general aspects of climate, season, soil, and crop, etc., the details in each particular case have to be accurately determined so that the greatest advantage is gained from an area for the volume of water that might be available. Such questions of irrigation water are almost always associated with problems of domestic water supply and with the subject of hydro-electric power generation. In the case of hydro-electric power from a storage reservoir the water emerging from the turbines returns to the stream and may be utilized for irrigation or domestic supplies.

Again, taking a wide view, it may be calculated that for a population of 2,000,000,000 people, now existing in the world, the consumption of water per day for drinking, washing, and domestic uses averages 10 to 15 gallons per head. In some industrial centres and large towns as much as 100 gallons a day per head may be used, while in desert and backward areas the consumption may be less than 5 gallons per head a day. Estimated at the average rate, the water drawn from rivers, wells, and piped sources amounts to about $12 \cdot 5$ cubic miles (roughly 50,000,000,000 tons) annually, reckoning 224 gallons or 36 cubic feet of water to the ton. In the employment of water for irrigation, or for domestic purposes, the supply is consumed and, so to say, lost, but in the majority of hydro-electric schemes the tail water is discharged back into the rivers and is again available for irrigation or town supplies, or both. The solution of most water supply problems is not the storage of the water but its cheap conveyance, by pipe and pumping, to the place where it is to be utilized. This introduces the subject of electrical power, which is perhaps most cheaply obtained from waterfalls where large perennial rivers drop into deep gorges. In Norway and Sweden, before 1939, hydro-electric energy was obtainable at certain sites at less than $0 \cdot 03d.$ per unit (kilowatt-hour), and the famous Niagara Falls were quoted as yielding electricity at about $0 \cdot 11d.$ per unit. More recent particulars are available of an agricultural area in Sweden of roughly 200 square miles (with a population of 30,000 people) which was supplied with about 1,000,000 kilowatt-hours (34 units per head) at a charge of $3 \cdot 5d.$ per unit. At the time of the Weir Report, 1925, Great Britain had a generator capacity of over 3,000,000 kilowatts and consumed (by sale) more than 4,000,000,000 units a year at a cost of $2 \cdot 047d.$ It was then anticipated that by 1940 the generating capacity in this country would be 10,000,000 kilowatts, the annual (sold) consumption nearly 21,400,000,000 units at a price of $1d.$ To-day

(1949) the price of electricity to domestic consumers in the London District area is from 1·85d. (winter cooking) to 0·65 (summer water heating) per unit (kilowatt-hour).

Approximate estimates of the world resources of hydro-electric power are about 250,000,000 kilowatts, but the quantity utilized (largely in the United States of America and Canada, but appreciably also in France and Italy) is of the order of 30,000,000 kilowatts. This total of hydro-electric power is less than half the total electrical energy used in all countries at the present time. The total is roughly estimated at 65,000,000 kilowatts, and it is probable that the electricity generated from steam-driven (coal-fired) stations exceeds that derived from water power. In the United States probably 66 per cent of the total electric power is obtained from the steam turbine and most of the remainder from water. It is of interest to mention some of the great hydro-electric power stations in the United States: in the Tennessee Valley Administration (Muscle Shoals or Wilson Dam, 184,000 kilowatts; Norris Dam, 100,000 kilowatts; Wheeler Dam, 64,000 kilowatts; Guntersville Dam, 72,000 kilowatts, etc.) the total expectation is for a generating capacity of 2,000,000 kilowatts. Niagara Falls (U.S.A.) has a capacity of 320,000 kilowatts; the Bonneville project about 450,000 kilowatts; the Boulder Dam scheme, 1,320,000 kilowatts; the Grand Coulee project, 1,940,000 kilowatts, besides other Government or State hydro-electric stations either under construction or being investigated. With these may also be mentioned some of the great steam-driven turbine generating stations in the United States, such as those of the Chicago district (Waukegan, 290,000 kilowatts; Crawford, 390,000 kilowatts; Powerton, 215,000 kilowatts; State Line, 358,000 kilowatts, etc.) and other regions where coal or oil or gas is used as fuel.

In the case of the steam-driven generating stations it is of interest to remember that although the normal steam pressures are from 300 to 600 lb. per square inch these pressures tend to go up to 1,200 and even 3,600 lb. per square inch in the more modern and larger turbogenerators. The pressure is accompanied by higher and higher temperatures up to 700° (with 200° superheat). Indeed, the temperatures are near the critical temperature of water (374° centigrade). Some of the generators are of great size, for example, the giant turbine in the Chicago State Line station, which has a capacity of 220,000 kilowatts. The large machine at the Richmond (Philadelphia) station has a generator of 165,000 kilowatts and is also a giant electricity producer. The limit to these steam-driven plants is not in the turbines but in the electrical heat loss in the generator itself. The introduction of better insulation and ventilation has been followed by the use of hydrogen for cooling which has permitted the giant type of electricity generator. However, a primary consideration in all large steam generator stations is the provision of a plentiful supply of cold water for the condensers (to maintain a high vacuum and so improve the thermal efficiency of the plant). In the case of the Richmond (Philadelphia) generating station, which is on the Delaware River and generates up to 285,000 kilowatts, the condensing water is drawn through concrete intake tunnels at the rate of 78,000 gallons per minute. After circulation the water is delivered back into the Delaware River a little warmer but otherwise unaffected. The amount of water in this case, 112,320,000 gallons (over 50,000 tons or 1,800,000 cubic feet) per day, would provide a supply of 25 gallons a day per head for a population of 12,000 people for a year.

From what has been stated in the previous paragraphs it is seen that water plays a fundamental role in the supply of cheap electricity. In the case of hydro-electric power it is the direct source of energy. In the other case of steam-driven generators, water is the transmission agent between the heat from the coal and the electric energy. In both cases there is continuous improvement in the plant and mode of transmission. Perhaps the greatest advances have been effected in the economy of the fuel. It is now a commonplace occurrence for 1 kilogram (2·2 lb.) of coal to yield 1 unit of electrical energy (a kilowatt-hour) and to show a thermal efficiency of over 20 per cent on the heat value of the coal. Even as far back as 1927 a report by the United States Geological Survey stated that " . . . the operators of the public utility power-plants performed the remarkable feat of generating $2\frac{1}{2}$ billion (thousand million) *more* kilowatt-hours of electricity with the use of 150,000 tons *less* of fuel than in the previous year ". However great the skill of the engineer, naturally it can never equal ideal resources, and for this reason countries endowed with large perennial streams with descents over falls or with elevated lakes in mountains subject to steady and adequate rainfall may have excellent resources of water power. These advantages appear in the case of Norway, Canada, and Switzerland, where hydro-electric power is cheaply harnessed. In these countries the *per capita* consumption of electric energy is, respectively, 2,000 kilowatt-hours, 1,150 units, and 990 units. The smallest expenditure on pipes and turbines is usually where high " heads " of pressure can be secured with relatively small quantities of water. Perhaps the record in this respect is Lac Fully, Switzerland, which provides a " head " of 5,412 feet on the turbines (2,260 lb. or roughly a ton per square inch pressure).

After this brief introduction to the subject of water it is hoped that the reader will be encouraged to delve further into this treatise, and by studying the properties of water and the art of utilizing this remarkable and yet everyday material, will find in it interest and information of value. There are many surprising physical and chemical characteristics possessed by water which have wide-reaching effects on natural phenomena, including such questions as to why deep ponds seldom freeze ; why frosts may occur in a valley and not affect the higher adjacent slopes ; how fresh water may occur along the coast below sea-level ; how corals may grow off the west coast of India and not off the east coast of Africa, and a host of other problems depending on wind and tide and the properties of water. The engineering aspects of water supply—storage, silting, leakage, etc., are well known, but even in these details much depends on the experience and knowledge of the engineer on the stability of hill slopes, the structure of the strata, the nature of the rocks, and other features which are best appreciated by those who observe and study the processes of nature and understand the movements of water underground as well as on the surface. There is a difference between the formation of cumulus clouds on a sunny day and the cloud that mantles an active volcano—the water vapour in the cumulus cloud perhaps arose from the sea a few hours previously, and the volcanic vapour returned to the atmosphere after millions of years' travel through the earth's crust.

PART I

THE NATURAL HISTORY OF WATER

PART I.—THE NATURAL HISTORY OF WATER

CHAPTER I.—THE CONSTITUTION OF WATER

INTRODUCTORY REMARKS

Geologists believe that the life of the earth, both of plants and animals, originated in the sea. This is in accord with the fact that neither plants nor animals can exist without water in some form or another. A study of the rocks has shown that the earliest sedimentary formations were laid down in water 1,000,000,000 years ago in the Eozoic era (Algonkian times). Although the strata have not yielded recognizable forms of plant or animal life (fossils), there is abundant evidence of marine animal life in the early (Cambrian) Palæozoic era of 500,000,000 years ago. Land plants do not become evident until the Silurian period of 350,000,000 years ago, and forests, such as might have yielded the plant matter for the formation of the coal in the Old Red Sandstone (of Bear Island, between Spitzbergen and Norway), do not become evident in the geological record until Devonian times (300,000,000 years ago). Whatever view is adopted for the formation of coal, whether the *in situ* theory of dismal swamps or the *drift theory* of transported vegetable debris, all are agreed that coal seams were formed under water. The entire history of the geological record of the life of the earth is based on the study of the animal and plant remains (fossils) which are found in the successive formations from the *Trilobites* of the Cambrian seas to the Berezov mammoth of the Yakutsk region in Siberia (found intact with flesh in edible state frozen in ice of the late glacial epoch of Northern Europe).

Coming to the human period, Noah's " Flood " is still an epoch of importance. According to the chronology of the English Bible, as calculated by Archbishop Ussher, in 1650, this " flood " occurred about 2400 B.C. in the drainage area of the Euphrates. There would appear to be, however, some error in the translation with regard to the word or name *Ararat*, which in the Hebrew can mean " sacred land " as distinct from the mountain, nearly 17,000 feet above sea-level, in Turkey, on the borders of Iran (Persia) and the U.S.S.R. (Armenia). A moment's consideration will show that the distribution of land and sea to-day is, respectively, 57,000,000 square miles of continental areas with an average height above sea-level of 2,500 feet, and 140,000,000 square miles of oceanic areas with a mean depth of 12,500 feet. If, therefore, Noah's flood raised the level of the sea 17,000 feet, to cover the top of Mount Ararat, the amount of water added to the oceans would need to be 1·33 times the volume already in the oceans. The problem, on the basis of the evaporation of sea water for rainfall on the land, would thus not only concern the source of the rainfall supply but the disposal of the run-off flow after the rainfall. Incidentally, it also means a rainfall of 16,250 feet in 960 hours, or roughly more than 200 inches an hour. These calculations support the opinion of many devout people that some elevated land in the Euphrates alluvial plain was meant as " the scared land " when the word Ararat was used in Genesis (chap. viii, 4).

From the days of Aristotle, between 500 and 400 B.C., to the beginning of the Christian era, water was grouped as one of the four " elements "—air, earth, fire, and water—which were recognized then as in a *setting* of ether. They were recognized as interchangeable, constitutionally, in respect that air conveyed a sense of " wetness and hotness ", earth referred to conditions of " dryness and coldness ", fire embodied the idea of "dryness and hotness ", and water conveys the impression of " wetness and coldness ". These early scientific theories are of considerable interest. There was a tendency at the close of the last century to treat these older ideas as bordering on absurdity. They were regarded as belonging to the same category as the alchemists' endeavours to transmute lead into gold or to find the elixir of life. The discoveries of the past half century, particularly the last twenty-five years, have not made those " theories " and " dreams " more incredible but have even proved them probable. Our own conceptions in physical chemistry have had to be readjusted. We now know that the two *laws*—relating to the indestructibility of matter and the conservation of energy—are two *phases* of a single principle. The atomic bomb has demonstrated that matter can be annihilated into energy, and the atom-smashing physicist, by nuclear fission, has transmuted several of the so-called elements from one kind into another.

Forms of Water

It is almost unnecessary to say that the pure, colourless, transparent, odourless, and almost tasteless liquid which occurs in our rivers and falls as rain is water. *Rain*, then, is typical water. It is liquid and replenishes the supply of water on the land, and it is this fresh or sweet water that is of the utmost importance to mankind. It is true the oceans and seas contain the largest quantities of water, but it is salt water, and as such undrinkable and useless for irrigation. When the liquid form, water, is cooled to 0° C. it is frozen solid and becomes *ice*. When water is heated to 100° C. it is transformed into the gaseous form known as *steam*. Water may be transformed into vapour below 100° C., as when it is evaporated by the sun's heat, and remain invisible as water vapour until the air pressure is reduced or the air containing the moisture is cooled. The moisture then forms droplets which remain suspended in the air but become visible as *cloud*, the typical flat-based cumulus clouds of sunny afternoons. Similarly, steam becomes visible and the water vapour is seen when a whistle blows or a boiler is blown. When the droplets of water in a cloud are further cooled drops form and the water falls as *rain*, usually from the dark, storm or nimbus cloud. When a moisture-laden cloud drives on to a hillside or comes in contact with the ground the name given to the misty air is *mist*. A *fog* develops when a mist produces a condition of considerable obscurity, but it is recognized that fogs are not caused by mist (droplets of water suspended in the air) alone. Dust and particularly smoke particles in combination with a normal mist produce the thickest fogs in London.

When air carrying invisible water vapour is brought against a cold surface the moisture (droplets and drops) is deposited on the cold surface as *dew*. If the cold surface is at a freezing temperature the moisture is deposited as ice crystals and the deposit is known as *frost*, or more commonly as *hoar-frost*. Where the water vapour in the air is chilled below the freezing point the moisture forms flakes of ice crystals which float down as *snow*. In those cases where the fall is a mixture of rain and snow the term *sleet* is applied. Where a rain cloud is involved in a thunderstorm and the drops of rain are carried upwards and frozen and fall as pellets of ice the name hailstone or *hail* is used for the ice pellets. In the higher mountains and the polar regions, the snow becomes compressed into streams and sheets of ice, which flow down the valleys as glaciers or gravitate to the sea as continental ice-sheets, and break off and float away as icebergs and icefloes. Although the normal condition of ice is brittleness, these masses in glaciers and continental sheets show marked fluidity under pressure, possibly involving the pheonomenon of regelation. These considerations of melting point for ice and vaporization into steam for water, and the fact that ice is a true crystalline form of water, which is of a definite chemical composition, introduce the question of whether water may not be regarded as a mineral, in the sense that mercury is a mineral. There can be no question that, geologically speaking, water is a mineral and that ice must be regarded as a rock.

Mode of Occurrence

The principle mode of occurrence of water on the earth is in the liquid form. It covers 140,000,000 square miles of the earth's surface to an average depth of 12,500 feet as the hydrosphere or oceanic areas. All this vast quantity of water is salt, carrying roughly 3·5 per cent of dissolved salts, most of these (75 per cent nearly) represented by common salt (sodium chloride, NaCl). By comparison with the oceanic waters the river and lake supplies of fresh water are very small, perhaps not enough to raise the level of the oceans even 1 foot (more accurately about 9 inches). A much greater supply of fresh water occurs in the form of ice and snow in the polar regions and in the higher mountain ranges. The quantity thus locked up, but which adds to the rivers, though in turn replenished by snow falls, would, if melted and added to the oceans, raise the present sea-level 30·4 feet. The water which percolates into the rocks and is held in the pore spaces and fissures, and emerges in wells and springs, is also considerable. It is computed that under the land, and down to the level of the sea, the porosity is 4 per cent of the total rock volume, and that the water thus held below ground level is greater than that in the ice caps and glaciers. If made available to the ocean, this underground water would raise the level of the oceans about 41·8 feet. The commonest source of supply is from rain, from the water vapour in the atmosphere, and it is a remarkable fact that normal air contains little water.

It is generally true to say that half the total water in the atmosphere lies within a height of 7,000 feet above sea-level, and that three-quarters of the atmospheric water occurs below a height of 12,000 feet. The "homogeneous" atmosphere is taken as extending to a height of 26,000 feet above sea-level. Assuming an average of 3 grains of water per cubic foot of normal air (564·90 grains per cubic foot), the quantity represents about 1·65 inches difference in the level of the ocean. Taken at 8 grains the amount would only raise the level of the oceans 4·4 inches if all the moisture held by the atmosphere was condensed. This, however, takes no account of the water vapour which is carried by the air as a result of the sun's evaporation from the oceans and precipitated on the land as rain and snow annually. The water vapour held in the atmosphere serves another purpose, while the atmosphere itself, the air, is a great carrier of water, as moisture, from the oceans to the land. This is best estimated as rain and snowfall, say an average of 30 inches from a surface of 100,000,000 square miles (in terms of evaporation). Much of this falls as rain on the land, but a great deal falls back into the sea or comes as snow in the polar regions. We may presume a rainfall on the land of about 30 inches for an area of about 50,000,000 square miles.

The above are the main sources of water, fresh or sweet, for the land. The oceans furnish the source from which the sun evaporates moisture, and the air, also under the influence of the sun's heat, carries to the land the moisture that is precipitated on it as rain and snow. This is the great circulation system which keeps the lakes stored and rivers flowing with water. It also supplies the water to the underground store from which springs, wells, and artesian aquifers are replenished. In addition it is the rain which waters the plants and forests and field crops in regions where there is no need for special irrigation systems. It is the infiltrating water from rainfall which reacts on rocks and minerals in the process of hydration and weathering, and becomes combined or held very strongly in the kaolinized, or lateritized, or otherwise hydrated rock. Materials such as laterite may hold up to 25 per cent of their weight of combined water, although the original rock may have had a water content of less than 1 per cent.

The Physical Properties of Water

The chief physical properties of water (as liquid), under normal conditions of temperature and pressure, are (i) its specific gravity or density ; (ii) specific heat ; (iii) boiling point ; (iv) freezing point ; (v) viscosity ; (vi) surface tension ; (vii) latent heat of vaporization ; (viii) coefficient of volume expansion ; (ix) thermal conductivity ; (x) electrical resistance ; (xi) index of refraction ; (xii) dielectric constant, etc.

(i) *Density* : Water has its smallest volume or maximum weight at 4° C. under a pressure of 760 mm. (atmospheric pressure). At 3·98° C. one cubic centimetre of water weighs one gramme, and at 0° C. the same volume of water weighs 0·99987 grammes. The density or specific gravity of normal sea water at 0° C. is 1·025 and ice about 0·9, so that fresh water can float on salt water, and ice will float on fresh water (even at boiling temperature). Water is used as the standard of density.

(ii) *Specific heat* : The amount of heat required to increase the temperature of one cubic centimetre of water 1° C. is the unit of heat known as the calorie. If the specific heat of water is taken as unity, that of most rocks, granite, marble, coal, etc., vary from 0·21 to 0·19. Sea water has a specific heat of 0·94 and ice has 0·502, while the metals have still lower specific heats—aluminium 0·22, iron 0·117, zinc 0·0935, lead 0·0305, etc. Alcohol has a specific heat of 0·6, benzene 0·41, while steam averages 0·488, helium 1·25, and hydrogen 3·402, with nitrogen 0·235 and oxygen 0·242 only.

(iii) *Boiling point* : The boiling point of water is 100° C. and normal sea water boils at about 104° C., mercury at 356·7° C. Olive oil boils at about 300° C., turpentine at 159° C., alcohol at 78·3° C., and ether at only 34·6° C. Sulphur melts at 115° C., tin at 232° C., and lead at 327° C.

(iv) *Freezing point* : Water becomes ice at 0° C. while sea water freezes at about − 9° C. Benzene freezes at 5·4° C., turpentine at − 10° C., mercury at − 38·8° C., alcohol at − 115° C.

Among the other physical properties of water are :

(v) *Viscosity* : Water is used as the standard of viscosity : it has a viscosity of unity at 20° C., and varies as follows—at 0° C. its viscosity is 1·791 ; at 10° C., 1·3077 ; at 20° C., 1·005 ; at 25° C., 0·895 ; at 50° C., 0·549, and at 100° C., 0·2338. The unit is known as a centipoise (equals 10 millipoise). The viscosity constants of other liquids (at

25° C.) are—benzene 0·649, kerosene 2·375, spindle oil 92·0, and castor oil 620 (compared with water about 0·9). While temperature reduces the viscosity of water, this property is anomalously affected by pressure, decreasing with pressure when the temperature is below 30° C., and vice versa. This raises an interesting problem in water movement at great depths.

(vi) *Surface tension* : This is the phenomenon of globules of water on a greasy surface or very hot surface and of the experiment of floating a slightly greased needle on water. It suggests a *skin* on the surface of water. While viscosity is measured in dynes per square centimetre, surface tension is determined in dynes per centimetre (with air at 20° C. as the basis of measurement). With the surface tension of water as 73·5 dynes per centimetre, that of mercury is 520, and alcohol is only 21·7. The corresponding viscosities are (in dynes per square centimetre per unit of velocity gradient 0·01006, 0·0156, and 0·0119 respectively, for water, mercury, and alcohol. Surface tension also affects the rate of flow of water through the interstices and pore spaces in the rocks. This is why clays, although with large porosity (of volume) practically prevent the infiltration of water through them (while absorbing and holding considerable amounts of water, and even becoming fluid with the quantity taken up).

With regard to the other physical properties :

(vii) The *latent heat of vaporization of water*. It is well known that to convert one cubic centimetre of water at 100° C. into steam at the same temperature, under atmospheric pressure (760 mm. of the mercury barometer), requires the addition of 538 calories. This is the latent heat in steam, which must be given up whenever steam is condensed back into water. The volume of steam generated is roughly 1,600 times that of the water used. The specific heat of steam is 0·488, where that of air is 0·237 and oxygen is 0·242, while the density of steam is 0·581, air is 1·293 and oxygen is 1·429 in grammes per litre (61 cubic inches equal 1 litre of 1,000 c.c.s). The critical temperature of steam is 374° C. and its critical pressure is 217·5 atmospheres (about 3,107 lb. per square inch). In this connection may be mentioned the *latent heat of fusion* or liquefaction of *ice*, which is the amount of heat to be added to ice at 0° C. to convert it into water at the same temperature (under atmospheric pressure). The amount is 79·77 calories, so that to convert one gramme of ice into steam, from 0° C. to 100° C., requires 79·77 plus 100 plus 538 or a total of 717·77 calories under atmospheric pressure. The density (specific gravity) of ice is about 0·9 (water equal to 1·0 at 4° C.), its specific heat is 0·502 (water 1).

(viii) The coefficient of *volume expansion* of water is taken as 0·00015 per degree centigrade, compared with 0·00110 for alcohol and

0·000182 for mercury (the two liquids used in the making of most thermometers).

(ix) The *thermal conductivity* of water is 0·00136, compared with 0·000423 for alcohol, 0·0197 for mercury, 0·485 for aluminium, 0·912 for copper, 0·0000522 for air, 0·0000524 for nitrogen, and 0·000318 for hydrogen.

(x) The *electrical resistance* of water (ohms per cubic centimetre at 18° C.) is $9·1 \times 10^6$, while that of mercury is 0·0000958, aluminium is 0·000003, and copper is 0·0000016 (or 1·6 microhms). The electrical resistance of ice is 3×10^8, and quartz is 1×10^8 and marble and slate much greater, considered as insulators.

There remain—

(xi) The *index of refraction* of water which is 1·333 for pure water and 1·343 for sea water, 1·362 for alcohol, 1·504 for benzene, and 1·632 for carbon disulphide. The refractive index of ice is 1·31, quartz is 1·544, and diamond 2·417.

(xii) *Dielectric constant* (K) and dielectric strength (volts per millimetre) are important properties. The former, or, as it is sometimes called, specific inductive capacity of water, is 80, where ice is 3 and air is 1·0006, hydrogen 1·00026, alcohol 26, benzene 4·37, glass from 5 to 10, and mica 6 (varies from 5·7 to 6·7). The *coefficient of volume expansion* of water (between 10° and 20° C.) is 0·00015 compared with 0·0011 for alcohol, 0·00124 for benzene, and 0·000182 for mercury. The linear expansion of solids compares as follows : ice 0·51, lead 0·2924, silver 0·1921, copper 0·1678, and graphite 0·0786. *Velocity of sound* in water is 4,714 feet (1,437 metres) per second as compared with 1,087 feet per second in air, 4,193 in hydrogen, 11,670 in copper, and 16,820 in iron, and up to 20,000 in some types of glass. The pressure of water vapour affects the velocity of sound in the air by increasing it as the humidity increases, which it would do by an increase of temperature, but the difference in the velocity of sound in dry air (1,087 feet per second) and warm air of high humidity is not great.

Since the atmosphere is the channel through which the *water vapour* obtained from the ocean is carried to the land and precipitated as rain and snow, it is important to understand the behaviour of the water vapour in transit. The table below shows that water vapour exerts a pressure of 4·58 mm. (of mercury) at 0° C. (and each cubic foot of air at that temperature can carry 2·37 grains of water ; at 25° C. it exerts a pressure of 23·76 mm. and carries about 12 grains of water per cubic foot of air ; at 50° C. it exerts a pressure of 92·54 mm. and a cubic foot of air may contain 36 grains of water ; at 100° C. it exerts a pressure of 760 mm. (equal to the atmosphere) and is all steam (about 254 grains per cubic foot, or 0·581 grammes per litre) :—

Temperature. °C.	Vapour Pressure. mm. (mercury).	Water Vapour (weight). Grains per cu. ft.
− 10	2·16	
− 5	3·17	
0	4·58	2·37
10	9·21	4·28
15	12·79	
20	17·54	8·20
25	23·76	12·10
30	31·83	14·00
35	42·19	
40	55·34	
50	92·54	36·00
60	149·50	
70	233·80	
80	355·50	
90	526·00	
100	760·00	253·93
150	3,569·00	
200	11,647·00	

The term vapour tension is often used in place of vapour pressure for the fraction of the atmospheric pressure which is exerted by any moisture or water vapour in the air. The amount of this moisture is measured by a hygrometer and expressed as the number of grains per cubic foot of the atmosphere or in terms of inches of mercury (1 inch is equal to 2·54 mm., and 15·43 grains equal 1 gramme). These measurements give the *absolute humidity* of the air. The saturation of air (with water vapour) is entirely dependent on the temperature of the air, but as the air is seldom still, invariably rising or falling, the air is either unsaturated or precipitating moisture (as rain or snow). The term *relative humidity* is therefore applied to the water vapour that would saturate the air at any particular temperature. Thus, the absolute humidity of the air might be read on the hygrometer as 8·2 grains per cubic foot at a temperature of 30° C. where the humidity (saturation) can be 14·00 grains of water. Air of this character will be dry and capable of taking up more than 5 grains of water per cubic foot. A wind (light breeze) of 15 miles an hour having the above temperature and humidity will permit the air within 10 feet of the ground and across a width of a mile to take up 300,000 gallons of water (as water vapour by evaporation) from an area of a mile by 15 miles within an hour. This means only 20,000 gallons per square mile per hour, which is negligible when compared with 1 inch of rainfall on a square mile (equals 14,500,000 gallons).

Analyses of the air are usually shown as *dry air* :—

Nitrogen . .	78·122 per cent by volume	75·539 per cent by weight
Oxygen . .	20·941 ,, ,, ,,	23·024 ,, ,, ,,
Argon . .	0·937 ,, ,, ,,	1·437 ,, ,, ,,

With the argon there occur the rare gases krypton, xenon, helium, and neon in the proportions 0·028, 0·005, 0·0004, and 0·00123 per cent by volume. In addition to these the air contains perhaps 0·01 per cent hydrogen, 0·033 per cent carbon dioxide, and 1·40 per cent water vapour by volume. Of these other constituents, which also include ammonia, ozone, nitric acid, sulphur, organic matter, etc., in traces as well as finely suspended solids, there is no question that the water value is next in importance to the oxygen itself. In that wonderful treatise "The Data of Geochemistry" (*Bulletin 770*, 1924, U.S. Geological Survey), Dr. F. W. Clarke (p. 58) has briefly discussed the primitive atmosphere. He states that "If we accept the nebular hypothesis, we are likely to conclude that the atmosphere is merely a residuum of uncombined gases which were left behind when the globe assumed its solid form. That seems to be the prevalent opinion, although it must be modified by the observed facts of volcanism. The outer envelope of the earth receives reinforcements from within . . .". The planetismal hypothesis put forward by T. C. Chamberlin is also considered as follows : ". . . a planet built up by slow aggregations of small, solid bodies. Each of these particles, or meteorites, carries with it entangled or occluded atmospheric material. In time the accumulation of originally cold matter developed pressure enough to raise the central portions of the mass to a high temperature, and gases were then expelled. Thus the atmosphere was generated from within the globe instead of remaining as a residuum around it. . . ." Whatever may be the true origin of the atmosphere there is very little doubt that large volumes of water are given out from volcanoes, but it is a question whether this is greater in quantity than the water absorbed by the processes of rock decomposition and hydration.

Dr. Harold Jeffreys has made a comment on this subject of the origin of the seas. He wrote (*Nature*, vol. 114, 1924, p. 934) : "It is probable that most of the ocean has come out of the earth in geological time. A fact noticed by Aston supports this view. The amounts of the inert gases in the earth and the atmosphere together are of the order of a millionth of those of even the rarer of the other elements. This is explained at once if the earth was originally heated ; for these gases, being unable to form compounds, would have no possible resting place except in the atmosphere and would be lost by diffusion into space. The materials of the present atmosphere and ocean, being volatile and having molecular weights less than those of krypton and xenon, would also be lost, and must therefore have been enclosed in compounds within the earth."

Chemical Characteristics of Water

The oceans constitute the great source of water on the earth's surface, where it is perhaps the most abundant substance visible. Pure water is a compound of hydrogen and oxygen in the proportion of two volumes of the former to one of the latter, or 2·016 parts by weight of hydrogen to 16·00 of oxygen. The chemical formula of water is written H_2O. It occurs in nature in the solid, liquid, and gaseous forms as ice and snow, water, and water vapour or steam respectively. Water has been used as the standard of measurement of heat (the calorie) and volume (the litre), and for comparison as for specific gravity (density), relative viscosity, etc. It is essential to all plant and animal life, and enters into combination with mineral (inorganic substances). The oceanic or sea waters contain 3·5 per cent of dissolved salts, chiefly sodium chloride (NaCl, 2·7 per cent), magnesium chloride (MgCl, 0·4 per cent), magnesium sulphate ($MgSO_4$, 0·2 per cent), calcium sulphate ($CaSO_4$, 0·15 per cent), and potassium chloride (KCl, 0·05 per cent). With 3·5 per cent of dissolved solids the salinity of sea water is reckoned at 35 (parts per thousand), but it is very difficult to measure salinity directly as the density of sea water is determinable (and depends on temperature and salinity) by the hydrometer at a standard temperature (32° or 60° F. or 0° and 15° C.). A density of 1·0 means zero salinity, density 1·0138, salinity 20·0, density 1·0260, salinity 35·0, etc., at a temperature of 60° F.

From the time rain falls on the ground and until it flows back into the ocean the water is at work. It brings with it some of the impurities in the air—carbon dioxide, acids, etc., and with these water dissolves soluble components from the soil and strata, or reacts and decomposes the rock minerals as it percolates below ground. However, it is recognized that of the rainfall, a part is re-evaporated, a part is held by the soil and plants, some runs off into the rivers, and the remainder sinks into the ground. It has been estimated by Sir John Murray (*Scottish Geographical Magazine*, vol. 3, 1887, p. 65) that of a total annual rainfall of 29,347·4 cubic miles, no less than 6,524 cubic miles run-off through the rivers to the sea and carry, in solution alone, 2,735,000,000 tons of dissolved solids. (One cubic mile of river water weighs 4,205,650,000 tons, roughly, and carries about 420,000 tons of foreign matter on an average.) In his book *The Realm of Nature*, Hugh Robert Mill gives (pp. 178–9, 1932 ed.) average analyses of the salts in river and sea water, pointing out that " The water of the ocean contains nearly 200 times as much dissolved solids as the water of rivers. Sea water, indeed, is at once recognized by taste as *salt*, while river water is pronounced *fresh*. The analyses are as follows :—

Salts in River-water.	Per cent.	Salts in Sea-water.	Per cent.
Calcium carbonate	42·90	Sodium chloride	77·70
Magnesium carbonate	14·80	Magnesium chloride	10·80
Silica	9·90	Magnesium sulphate	4·70
Calcium sulphate	4·50	Calcium sulphate	3·60
Sodium sulphate	4·20	Potassium sulphate	2·50
Potassium sulphate	2·70	Calcium and magnesium carbonate	0·30
Sodium nitrate	3·50	Magnesium bromide	0·20
Sodium chloride	2·20	Other salts	0·20
Alumina and Iron oxide	3·60		
Other salts	1·30		
Organic substances	10·40		
Total	100·00	Total	100·00

Thus, it is seen that while the salinity (parts per 1,000) of sea water may average 35, that of river water may be less than 0·175. Also that the salts present (dissolved) in the two waters may be different. Sea water contains 2·71 per cent of sodium chloride as against a mere trace in river water. It averages 0·378 per cent sulphates as against 0·057 per cent in river water, but river water carries 0·288 per cent of carbonates as against a trace (0·00105) in sea water. Similarly, river water has 0·05 per cent silica as against perhaps 0·0004 per cent in sea water. It is obvious that river waters will vary (in their dissolved solids) according to the rocks of the country they traverse. It is thought that some living creatures, plant and animal, in the sea remove the silica and carbonate brought in by the river waters. Although sea water is slightly alkaline it is still capable of dissolving carbonate of lime. This solubility increases under great pressure. (It has been calculated that in the deeps of the ocean the pressure may be as much as 4 tons to the square inch, and that owing to the slight compressibility of sea water 11,000 cubic feet are compressed into 10,000 cubic feet, and that if water was not compressible the ocean level would rise 200 feet.) It is doubtful whether pressure or temperature seriously affects the composition of sea water in the depths of the ocean.

In this connection, however, mention must be made of the confusion caused by the manner in which chemical analyses of water (of the dissolved solids in water) are prepared. Dr. F. W. Clarke (*Data of Geochemistry*, 1924 ed., p. 64) has drawn particular attention to this procedure as follows : " . . . a water is found to contain sodium, potassium, calcium, magnesium, chlorine, and the radicles of sulphuric and carbonic acids ; or, in ordinary parlance, three acids and four bases. If these are combined into salts at least twelve such compounds must be assumed, and there is no law by which their relative proportions can be calculated . . . each chemist allots the several acids to the several bases according to his individual judgment . . . all the chlorine may be assigned to the sodium and all the sulphuric acid to the lime, and the result is a meaningless chaos of assumptions and uncertainties. . . ." He adds : " Before proceeding further, it may be well to consider a single water analysis. . . . I will take W. P. Headden's analysis of water from the Platte River, near Greeley, Colorado. . . . " (*Bulletin*, Colorado Agricultural Station, No. 82, 1903, p. 56.) See top of page 12.

In the first column the results are given in oxides as in a mineral analysis. In the second column in terms of salts but recalculated in parts per million. The third column is given in accordance with the residue in radicles or ions and in percentages of the total anhydrous inorganic solids. So far as appearance goes, the above analyses might represent three different waters. In the interpretation of any water analysis the first question is its accuracy. When an analysis is stated

ANALYSIS OF WATER STATED IN DIFFERENT FORMS

Grains per Imperial Gallon.		Parts per Million.		Per cent.	
SiO_2	0·891	$CaSO_4$	457·7	SiO_2	1·26
SO_3	32·601	$MgSO_4$	236·0	SO_4	55·28
CO_2	4·554	K_2SO_4	9·4	CO_3	8·78
Cl	2·681	$NaSO_4$	62·5	Cl	3·79
Na_2O	11·463	NaCl	63·2	Na	12·02
K_2O	0·355	Na_2CO_3	156·9	K	0·41
CaO	13·117	Na_2SiO_3	21·9	Ca	13·24
MgO	5·530	$(FeAl)_2O_3$	2·7	Mg	4·69
$(FeAl)_2O_3$	0·189	Mn_2O_3	2·7	R_2O_3	0·53
Ignition	2·397	Ignition	34·2	Ignition omitted	
Mn_2O_3	0·189	Excess SiO_2	1·3		
	73·967		1,048·5		100·00
Less O − Cl	0·604			Salinity 1,014 parts per million.	
	73·363				

in terms of the radicles actually determined the value of its reliability is easier to decide.

The following analyses (also quoted from *Data of Geochemistry*) show the progressive increase in matter in solution :—

	(1.)	(2.)	(3.)	(4.)	(5.)	(6.)
CO_3	47·42	47·26	41·66	0·30	13·11	trace
SO_4	3·42	5·77	5·19	7·59	7·22	0·31
Cl	1·89	2·42	1·51	55·46	41·47	65·81
NO_3	0·86	0·38	—	—	—	—
Ca	22·42	22·33	20·08	1·21	10·67	4·73
Mg	5·35	6·52	4·52	3·79	4·88	13·28
Na	5·52	4·10	3·92	30·53	18·11	11·65
K	w.a.	w.a.	w.a.	1·12	1·14	1·85
SiO_2	12·76	11·16	23·12	—	w.b.	w.b.
Fe_2O_3	0·16	0·06	—	—	3·40	trace
Br	—	—	—	w.Cl.	—	2·37
	100·00	100·00	100·00	100·00	100·00	100·00
Salinity in parts per million	0·60	108·00	160·00	35,000	7,700	192,000

(1) Lake Superior, at Sault St. Marie, almost pure water.
(2) Lake Huron, Port Huron, slightly " hard " water.
(3) St. Lawrence River, at Ponte des Cascades, taken up silica.
(4) North Atlantic water from between Norway and Iceland.
(5) River Jordan, near Jericho, obviously " hard and saline ".
(6) Dead Sea, near north end, quite different to Atlantic.

These analyses show a pure lake water (1) ; a typical river water (3) ; an oceanic analysis (4) ; a contaminated river water (5), and an exceptional enclosed drainage lake water (6). It would be difficult to find a purer water than (1) nor easy to discover a brine resembling (6). The letters w.a. and w.b. mean " with above " or " with below " showing that the potassium is included with the sodium, or the silica with the ferric oxide. The salinity is given in parts per million, but this is a matter of choice since the solid in fresh waters is so small, while the amount in sea water and in brines and bitterns is usually large, it is customary to give the salinity of the latter in parts per 1,000 or 100.

There are many salts, such as the alums, which crystallize with water in combination or water of crystallization, in very appreciable amounts, as seen in the general formula $R'R''(SO_4)_2.12H_2O$. Such compounds of copper sulphate also retain water of crystallization when crystallized from an aqueous solution—$CuSO_4.5H_2O$. In many cases water is held in combination as water of hydration, such as in the slaking of lime and setting of plaster of Paris and of Portland cement. This may belong to some general process of hydration in nature, where anhydrate is converted into gypsum by taking up water in combination, or in the hydration of the feldspars, into zeolites (secondary minerals formed in some eruptive rocks from feldspar) such as heulandite $(CaAl_2Si_6O_{16}.5H_2O)$ at relatively high temperatures, or into Kaolinite $(Al_2Si_2O_7.2H_2O)$ by hydrous alteration at ordinary or moderate temperatures during rock-decomposition below ground level. The water entering into combination with the hydrates of alumina and ferric oxide in laterite may represent as much as 24 per cent or more of combined water in, say, an aluminous laterite. These aspects of the reaction of water on rock forming minerals and salts (compound where one or more atom of hydrogen of an acid have been replaced by a metal atom or electro-positive radicle), distinct from simple solution, are of the greatest importance in the chemistry of the earth's crust. As is well known, water, or as it should be called, hydrogen oxide (H_2O), the liquid, forms when hydrogen and oxygen combine—the usual experiment being to burn hydrogen in oxygen and collect the product by cooling the gases formed.

It has been long claimed that the liquid, H_2O, contains also molecules of H_4O_2 and H_6O_3, etc., but the discovery of isotopes (substances having identical chemical properties but different atomic weights) has been followed by the discovery of isotopes of oxygen (16), both 17 and 18, in the proportion 2,500 : 5 : 1 of 16 : 18 : 17, and of hydrogen, named deuterium, in the proportion 7,000 : 1 : H : D. Consequently, the discovery of the compound D_2O and its separation from ordinary water, H_2O, by the electrical decomposition of water (hundreds of gallons have yielded a few millilitres of practically pure D_2O or, as it is commonly called, *heavy water*). Heavy Water has the same chemical properties as normal water, but it has different physical properties. It is about 10 per cent heavier (than water) ; freezes at 3·82° C. (about the temperature of maximum density of normal water) ; boils at 101·42° C., and has its maximum density at 11·6° C. Its concentration in lake water is about 1 in 6,000. It has solid phases corresponding to ice i, iii, v, and vi. The other characteristics are as follows :—

	Ordinary Water.	Heavy Water.
Dielectric constant	78·5 cgs e	80·7 cgs e
Viscosity coefficient	9·95 millipoise	10·99 millipoise
Ion product	$1·0 \times 10^{-14}$	$0·2 \times 10^{-14}$
Surface tension	72·0 dyne/cm.	72·0 dyne/cm.
Freezes at	0° C.	3·82° C.
Boils at	100° C.	101·42° C.
Density at 4° C.	1·0	1·0079

In general, salts are less soluble in D_2O than in H_2O and the rates of chemical reactions in solution are less in D_2O. Heavy water is used to slow the speed of neutrons in nuclear fission or atom-smashing investigations.

The relatively high dielectric constant of water is thought to explain its ionizing power and solvent action. It possesses sufficient ionizing properties to convert H_2O into H' or H_3O' and OH^- hydrions and hydroxylions respectively, and thus to cause the hydrolysis of a weak acid or base in its solutions: RX + H.OH into RH + X.OH. In the case of substances such as phosphoric chloride and aluminium sulphide complete hydrolysis occurs when they are dissolved in water: $PCl_3 + 3H_2O$ becomes $H_3PO_3 + 3HCl$, and $Al_2S_3 + 6H_2O$ yields $2Al(OH)_3 + 3H_2S$. This formation of an acid and a base from a salt by interaction (ionic dissociation of) with water is termed saponification when the hydrolysis of fats is concerned, but the interaction is the same in both cases. It is found that metallic carbides, such as calcium carbide, produce hydrocarbons with water: $CaC_2 + 2H_2O$ yields slaked lime, $Ca(OH)_2$, and acetylene, C_2H_2. Metallic nitrides normally yield ammonia, NH_3, with contact with water, and leave the metal as an hydroxide when the gas, ammonia, is liberated. Metallic hydrides give up their hydrogen in their reactions with water as the gas, hydrogen, escapes. Under certain conditions some metallic and non-metallic elements, e.g. iron and carbon respectively, can be directly oxidized by water (including steam) to form oxides and give up hydrogen. Hydrogen is produced by the reaction of steam and spongy iron at about 650° C. The cheapest source of hydrogen manufacture is by removing carbon monoxide from "water gas" as follows, $CO + H_2O \rightleftarrows CO_2 + H_2$, in the presence of iron oxide (with chromium oxide, Cr_2O_3, as catalyst). Some oxides, such as lime (quick lime), CaO, and certain hydrated oxides, such as $Al_2O_3.H_2O$, react with water to form basic and acid hydroxides—$Ca^+ + 2OH^-$ in the former example, and $Al(OH)^+_2 + OH^-$ in the latter case. Steam reacts with the halogen elements—chlorine, bromine, etc.—to liberate oxygen with a reaction as $2Cl_2 + 2H_2O$ gives $4HCl + O_2$. Reactions of steam with sulphur yield sulphur dioxide, SO_2, and sulphuretted hydrogen, $2H_2S$.

Heavy Water in Atomic Energy

It is outside the scope of this treatise to discuss the subject of atomic energy except in connection with the use of heavy water (already

Geological Survey of India

PLATE III.—THE BEGINNING OF A RIVER from the mouth of a Glacier. A view of Snowfields and Glaciers.

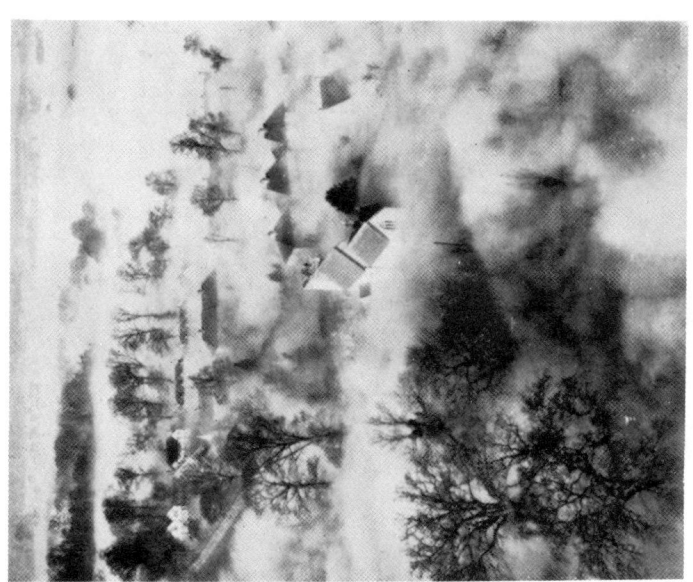

PLATE II.—WATER VAPOUR AS GROUND MIST in Hertfordshire, England.

[*Facing page 16*

PLATE IV.—AN ICEBERG MELTING IN THE NORTH ATLANTIC.

PLATE V.—THE FROZEN MAMMOTH IN SIBERIA.
This specimen was found in the ice on the Berezov tributary of the Kolymsk River, Yakutsk, Siberia, in 1900, and set up in St. Petersberg in 1903.

[*Facing page 17*

discussed above on p. 13). There are several publications dealing with the subject, none perhaps more authoritatively than *Atomic Energy : A general account of the development of methods of using atomic energy for military purposes under the auspices of the United States Government*, by H. D. Smyth (published in the U.S.A., reprinted by H.M. Stationery Office, London, 1945). It may be said, however, that the name " atomic bomb " has been given to the release of atomic energy (the conversion of matter into energy) with explosive violence as a result of nuclear fission (atom-smashing) in a " chain " or continued reaction. Lord Rutherford (1919) had demonstrated that, by bombarding nitrogen with high energy alpha particles a small amount of oxygen was produced ($He^4 + N^{14} \rightarrow O^{17} + H^7$; the helium is the a-particle. This was the first practical transmutation of elements—the alchemist's dream come true. However, it was not until 1932 that it was proved that neutrons (uncharged) particles are more suitable missiles for the transmutation of elements. In 1934 Irene Curie and F. Joliot discovered that some of the newly formed elements were unstable isotopes which emitted particles (or were radioactive). This phenomenon was called " artificial radioactivity ". In the same year Enrico Fermi showed that the most efficient projectiles for atom-smashing were neutrons which had been slowed down by elastic collisions with light atomic nuclei while passing through heavy water or other suitable substance. Fermi used his method for transmuting uranium (atomic number 92) into trans-uranium atom (numbers 93 and 94). In 1939, in January, O. Hahn and F. Strassmann announced that barium was one of the products when uranium was bombarded with neutrons.

The announcement that the recognition of barium (No. 56, Ba-137) was one of the products of uranium fission, probably with krypton (No. 36, Kr-83), was of very great importance since it showed that uranium had almost been split in half (not a mere fraction or so). This information was conveyed to Niels Bohr, in Copenhagen, by Lise Meitner and Frisch, refugees from Germany, and Bohr carried it to Washington where it was discussed fully with Einstein and others. Calculations showed that, theoretically, the loss of mass in such a violent transformation would be equivalent to 200,000,000 electron volts per single split atom (as against 4 electron volts when an atom of carbon combines with oxygen in a coal fire). It meant 50,000,000 times as much energy as was available from the same weight of coal, i.e. one pound of uranium would equal 9,000 tons of coal roughly. Fermi and others suggested that several neutrons might also be emitted and so the potentialities of a sustained or chain reaction were evident. It was soon appreciated that, with the fission of uranium (235) or plutonium with fast neutrons, the reaction would be uncontrolled and violently explosive ; with uranium (238) the normal effect would be that of an improperly or inefficiently fired explosive ; with " slow " neutrons both uranium and plutonium would be under control. The proportion of uranium 235 in normal uranium (238) was 1 in 140, and the latter was already a rare substance. By bombarding uranium 238 an unstable new element, neptunium (No. 93), was produced which, in emitting another electron, became the new trans-uranium element plutonium (No. 94). Great interest centred in the controlled chain reaction, but the difficulty here was to find a " moderator " which would permit bombardment with the " slow " neutrons. The chain reaction of the explosive type, atom-smashing uranium 235 or plutonium with " fast " neutrons, was simpler if the uranium 235 or the plutonium was made available (by bombardments with " slow " neutrons).

The subject of a suitable moderator seemed to be " heavy water ", and at Joliot's request the French Government secured the largest available quantity, about 165 litres (about 1·5 tons), from the Norsk Hydro Company of Norway. Before this could be utilized in Paris the German advance was closing on the capital and, with assistance from England, Halban and Kowarski escaped with it to Cambridge. By December, 1940, they were able to report that if a system of uranium oxide, or of uranium metal surrounded by " heavy water ", were sufficiently large, a sustained or controlled chain reaction was practicable. It was estimated that only a few tons of " heavy water " would be required if the uranium metal system were used. It is now thought that the German endeavour to build an atomic bomb was largely frustrated by two Anglo-Norwegian commando raids on the Norsk plant and its stock of " heavy water " damaged or made unavailable to the Germans. It cost the life of Leif Tronstad, professor of physical chemistry at Trondjem, who was a commando captain. In 1944 a site on the Ottawa river was fixed upon with the help of the Canadian Government for an atomic uranium and heavy water power plant. The heavy water was supplied by the American Government. It must be added, however, that although " heavy water " is a very satisfactory " moderator " for slowing the speed of fast neutrons, there. are other substances which are very efficient, for example, " heavy hydrogen " (deuterium) itself which was obtained in 1940 by H. C. Urey (Columbia University, New York). Among substances also are beryllium, graphite, etc. And, indeed, Fermi and Szilard recommended making a " cake " with the uranium in it as raisins embedded in the " moderator ".

Fermi's cake was the beginning of the so-called " pile ". The atomic pilot plant pile which was almost the first made was a 7 ft. cube of graphite with 7 tons of uranium oxide (as its raisins). It proved suitable for the self-sustaining chain reaction for the controlled release of atomic energy both for the production of industrial power and for the manufacture of plutonium (for use either for power or making an atomic bomb). The fact that there are substitutes for either heavy hydrogen (deuterium) or heavy water (D_2O) and the belief that atomic enegy may be liberated from other materials than uranium and plutonium by other methods than nuclear fission, suggests that there can be no secrecy as to the method of making the type of atomic bomb tested at Bikini Atoll. The following extract from the *Encyclopædia Britannica* (1947 ed.), vol. 2, p. 647D, under the heading " Peacetime Possibilities ", states : " . . . It is entirely within the range of possibility that some day methods of releasing atomic energy, other than fission, which will convert larger percentages of the mass involved into energy may be discovered. Should the day come when such nuclear reactions as are known to take place in the stars and the sun can be duplicated, it will be possible to drive an ocean liner across the Atlantic on the atomic energy contained in a glass of water. . . ." Another extract from a still more recent writer, Chapman Pincher (*Into the Atomic Age*, 1948, p. 134), states : " Scientists believe they know the processes involved in the sun's atomic chemistry. The main one seems to be the condensation of hydrogen atoms to form the gas helium. The union of hydrogen atoms in fours to produce single atoms of helium results in the disappearance of substance and the creation of a corresponding amount of energy. Actually the energy released from one pound of hydrogen by this process is seven times greater than the energy set free by the splitting of one pound of uranium atoms. If it could be repeated on earth under controlled conditions, then the raw material for atomic energy would be water, for hydrogen can be obtained from water simply by passing an electrical current through it. The hydrogen in half a pint of water would yield the power of about 600 tons of coal."

Chapter II.—The Distribution of Water

The water of the earth is distributed in the oceans and seas, in the lakes and rivers, in icefields and glaciers, in the soil and life of the earth, in the strata and rocks for a considerable depth below the surface. Rough estimates of the water in these parts of the earth were given in the Introduction of this treatise on p. xvi, but it is now necessary to examine the data more fully. This is best done by treating the earth's water as found in the atmosphere, hydrosphere, and lithosphere.

The Atmosphere

It is now commonly accepted that the atmosphere extends 200 miles upward or outward from the earth's surface, and it is considered in the following regions or layers : (i) the region of our weather, the troposphere, from the surface to a height of about 33,000 feet (6·25 miles or 10 kilometres to the tropopause) ; (ii) the region of constant ($-53°$ C.) temperature or stratosphere, from 34,000 feet to the upper air at a height of nearly 15 miles (nearly 80,000 feet) ; (iii) the ozonosphere, where ozone is formed by the sun's rays, from 15 miles to 25 miles or 40 kilometres ; and so on into (iv) the ionosphere, with its various layers, from 25 miles outward for another 200 miles perhaps. Climbers have got to within a few hundred feet of the top of Everest (29,002 feet), pilots have flown aeroplanes through the troposphere up to heights of over 10 miles (54,000 feet) in the stratosphere. Manned balloons have ascended to heights of nearly 14 miles (about 72,000 feet) into the ozonosphere. And instrumental data by other balloons have been obtained from still greater heights in the ozonosphere (almost to 20 miles). The abnormal propagation of sound (gunfire at great distances) has been ascribed to a temperature inversion ($-53°$ C. to perhaps $100°$ C.) of the attenuated atmosphere at a height of 50 kilometres or 32 miles. The phenomenon of the aurora borealis (and of aurora australis), due to electrical discharges in rarefied air, are regarded as proof of an atmosphere as high as 100 miles (160 kilometres). And the luminosity developed by meteors as they approach within 200 miles of the earth is regarded as evidence of the fringe of our atmosphere at that distance from the surface of the earth.

So far as problems of water or water vapour are concerned our interest is largely in the region of the troposphere. It extends from the ground where the air averages 4 per cent of water vapour (depending on the weather) to a height of 34,000 feet (the tropopause) above which no clouds occur. The highest cirrus clouds appear to lie at a height of 6 miles (roughly 32,000 feet). The occurrence of cirrus-cumulus clouds extends up to a height of 4 miles (roughly 21,000 feet), and the cumulus-nimbus clouds may form against highlands at heights of 1½ miles (8,000 feet). All these are direct evidence of water vapour in the atmosphere, but for rough calculations it has been customary to assume that half the moisture in the air occurs below a height of 7,000 feet, and to consider a homogeneous atmosphere as 26,500 feet thick. It is not to be concluded that water vapour is absent above the tropopause, since actual determinations in the stratosphere have been made (A. W. Brewer, 1945) and show that a relative humidity of 2 to 3 per cent may be proved by a frost-point hygrometer in middle latitudes. Irridescent or mother-of-pearl clouds at dawn and twilight at heights of 15 to 16 miles are regarded as evidence of higher humidities in the stratosphere (at its top and practically at the base of the ozonosphere).

Although the main constituents of atmospheric air are given in the proportions below :—

	By Weight.	By Volume.
Oxygen	23·024 per cent	20·941 per cent
Nitrogen	75·539 ,, ,,	78·122 ,, ,,
Argon, neon, etc.	1·437 ,, ,,	0·937 ,, ,,

and the argon component includes, by volume 0·028 per cent krypton, 0·005 per cent xenon, 0·00123 per cent neon, and 0·0004 per cent helium, there are several other constituents present. These are shown in a typical analysis of so-called pure air (below) and in an analysis of average air :—

By Volume.	Pure Dry Air. Per 1,000.	Average Air. Per 1,000.
Nitrogen	780·3	769·5
Oxygen	209·9	206·6
Argon	9·32	9·37
Neon	0·018	with argon
Helium	0·005	,, ,,
Krypton	0·001	,, ,,
Xenon	0·00009	,, ,,
Hydrogen	0·11	0·190
Carbon dioxide	0·30	0·336
Ozone in upper air		0·0015
Nitric acid		0·0005
Ammonia		0·008
Water vapour		14·00

The presence of dust particles is easily proved by looking across a sunbeam, and water vapour becomes evident on glass (which has been cold and is brought into a warm room) by the formation of dew. The amount of water vapour in the air is very variable and depends on the weather. Also, water vapour may become evident in the air as droplets of water, as ice in the form of snow, or be invisible as a gas until deposited as dew.

Although aqueous vapour is a variable component of the atmosphere it is the most important geologically. It dissolves and concentrates the other constituents of the air as it falls as rain. Although oxygen is perhaps the most active of the atmospheric gases it is assisted by moisture. Perfectly dry oxygen is practically inert and phosphorus would not burn readily in it, but the merest trace of moisture sets the combustion going strongly. Rain dissolves oxygen, nitrogen, carbon dioxide, etc., from the air, but the dissolved gases are in quite different proportions in water, and the percentage of oxygen may be as much as 34 per cent in air dissolved in water (rain). Temperature plays a very important part in fixing the amount of water vapour the air can carry, but this is also influenced by the pressure exerted by the water vapour in the air. The amount of water vapour in the air is measured by a hygrometer as relative humidity, but if the temperature falls the air might deposit moisture, and vice versa if the air warms up it may absorb moisture. These changes in the humidity of the air may be very great where mountains border a coastal plain in tropical regions. During the day a large amount of moisture may be taken up by the warm air, but as it rises against the mountains the air will be cooled and rain will be precipitated. If for saturated air a fall of $1°$ C. occurs for every rise of 500 feet, then a very marked condensation will be caused by a range of mountains 10,000 feet high. It is for this reason that a high mountain chain is an effective barrier to the transfer of moisture to the territories beyond.

In his book *Atoms in Action*, 1944, George Russell Harrison has an interesting chapter on " Outwitting the Weather " (p. 289). He wrote : " . . . three pounds of food and four pounds of water a day will keep the body functioning, but these would be of little use without 34 pounds of air daily. . . ." He points out that in winter the air brought into a warm room is dry and water might be added to the air (say, a gallon for a large room) with advantage to comfort. Similarly, due to heat in summer, air brought into a cool room will increase in relative humidity and cause lassitude. The effect is expressed as " It isn't the heat, it's the humidity ". In a crowded theatre during an exciting performance 1,000 persons will supply 18 gallons of moisture to the air of the auditorium, and the air-conditioning plant must be able to cope with this quantity per hour. He refers to the simple fact that a spray of cold water is able to give up or abstract moisture to or from the air, whether it is cool and dry or warm and humid, respectively. And mention is made of the introduction of refrigerator plant for air-conditioning the ventilation in deep mines, such as the gold mines of the Rand where the air temperature below ground may be 100° F. and the humidity 100 per cent. A single plant may be capable in a normal way of producing 2,000 tons of ice per day, but be employed in cooling and conditioning 150,000 cubic feet of air per minute for ventilation. He states that " It has always been easier to warm a room in winter than to cool it in summer, and cooling still costs from two to three times as much as heating. . . . Physical methods are best for producing cold ". All these examples are intended to show the importance of the moisture in the air to man.

Oscar E. Meinzer (" Physics of the Earth "—IX, 1942, *Hydrology*, 1st ed., 2nd imp.) includes among the valuable papers he has edited on the circulation of the water of the earth a contribution on Precipitation, by Merrill Bernand. In this the statement is made that " If all the moisture in the atmosphere were suddenly precipitated there would be enough to produce an average depth of 1 inch of water on the whole earth's surface ". Reckoning the earth's surface at 197,000,000 square miles, a homogeneous atmosphere about 5 miles (26,300 feet) deep and using the above thickness of water (at 4,089,000,000 tons of water to a cubic mile), the equation works out to about 1·3 grains of water per cubic foot of air as an average. This is on the basis of 7,000 grains to the pound (avoirdupois) and 147,197,952,000 cubic feet in a cubic mile. The water content appears on the small side, but a calculation taking 2·37 grains per cubic foot (air at 0° C. saturated with water can hold no more) and considering this for a height of 7,000 feet (up to the height below which half the moisture of the atmosphere is assumed to lie), the calculations give a result of also 1 inch thickness of water on the earth's surface (if all the moisture in the atmosphere was condensed). It seems to me that neither calculation allows enough for the moisture in air in the tropics where the temperature is higher and the moisture content must be greater. However, if an allowance of *3 grains* of moisture is allowed per cubic foot of air, as the average up to 7,000 feet, the total result is only 25 per cent greater and the layer of water would be nearer 1·25 inches than 1 inch thick on the earth's surface (if all the moisture in the atmosphere was condensed). It may therefore be safe to accept either figure or, say, an average of about 1·16 inches for the whole area of 197,000,000 square miles of the earth's surface. If the atmospheric water was all precipitated on the land (57,000,000 square miles) it would represent nearly 4 inches of rainfall. If it fell only on the oceans (140,000,000 square miles) it would raise the sea level about 1·64 inches. The total volume of the water in the atmosphere would be roughly 3,600 cubic miles, weighing 14,744,934,000,000 tons (on the average roughly 2·7 grains of moisture per cubic foot up to a height of 7,000 feet for half the total of the whole atmosphere).

The Oceans and the Seas

The general properties of sea water are shown below :—

Salinity in °/1,000.	0°	10°.	20°.	30°.	40°.
Freezing point	0°	− 0·53	− 1·07	− 1·63	− 2·2
Density for temperatures	1·00	1·0082	1·0161	1·0242	1·0323
	3·95° C.	1·86° C.	− 0·31° C.	− 2·47° C.	− 4·54° C.
Osmotic press. (atmos.)	—	6·4	13·0	19·7	26·6
Elev. of boiling point	—	0·16	0·31	0·47	0·64
Reduct. of vapour press. (mm.)	—	4·2	8·5	13·0	17·6
Specific heat	—	0·968	—	—	0·932
Viscosity for salinity of 35° (millipoise) 8·95 for fresh water (25° C.)		5·2 per cent greater at 0° C. and 4·0 per cent greater at 25° C. than fresh water at 25° C.			

Compressibility of sea water (35/1000) is given below :—

Depth in Metres.	Temperature in ° F.	Salinity 1/1000.	Density as found.	Specific Gravity without comp.
0	80·8	34·32	1·0222	1·0222
100	76·6	34·86	1·02376	1·02331
1,000	40·10	34·53	1·03215	1·02739
5,000	34·7	34·68	1·05107	1·02777
10,000	36·7	34·68	1·07123	1·02768

It may be added that the blue colour of the sea is due to the selective absorption of light in water. This is shown below for depths to 50 metres :—

Red	wavelength	650 uu	0·002/1000	absorbed
Yellow	,,	600 ,,	0·03/1000	,,
Green	,,	550 ,,	2·2/1000	,,
Blue	,,	450 ,,	201/1000	,,
Violet	,,	410 ,,	200/1000	,,

The following table shows the distribution of the oceans with the main continental lands :—

Oceanic Areas (millions sq. miles).		Land Areas (millions sq. miles).	
Pacific Ocean	67·50	Europe	3·90
Atlantic	34·30	Asia	17·40
Indian	27·60	Africa	11·60
Arctic	5·40	South America	6·85
Mediterranean	1·15	North America	9·55
Other seas	4·05	Australia	3·50
		Antarctica	4·20
Total oceans	140·00	Total land	57·00

The total given for the oceans is somewhat smaller and the total for the land areas is rather larger than is usually given in geographical books, but consideration is here given to the extensive " continental " shelf in the Atlantic and other oceans. For rough calculations it is permissible to use round numbers—57,000,000 or 58,000,000 square miles for the land and 139,000,000 or 140,000,000 square miles for the oceans. The average height of the land above sea level is 2,500 feet, with Mount Everest at 29,002 feet as the highest point. The mean depth of the oceans is taken as 12,500 feet with the Planet Deep, 35,410 feet, as the greatest depth.

A view of the earth (globe) with the Loire region of France as the centre of the hemisphere, includes the largest proportion of land to water, roughly 82,750,000 square miles of land to 114,250,000 square miles of ocean, while a view from the opposite side—with New Zealand as centre—covers the greatest spread of water, about 163,500,000 square miles, as against only 33,500,000 square miles of land areas. These proportions and estimates are not in agreement with many other estimates which have been published ; they are not intended as more than mere approximations of the distribution of land and water on the earth's surface. In many publications the figures are given in square and cubic kilometres. The conversion factor for converting square miles into square kilometres is obtained by multiplying the former by 2·59, and conversely the latter by 0·3861. To convert cubic miles to cubic kilometres, multiply by 4·165, and cubic kilometres by 0·2403 for obtaining cubic miles. For converting miles to kilometres (linear measures) multiply by 1·61, and vice versa by 0·625. The total surface area of the earth, 197,000,000 square miles, becomes 510,230,000 square kilometres (often given as 509,951,000), the oceanic area of 140,000,000 square miles calculates to 362,600,000 square kilometres (sometimes quoted as 361,059,000), and the land area of 57,000,000 square miles equals 147,630,000 square kilometres (which is also given as 148,892,000).

The following details of the oceans are quoted for general comparison from the *Encyclopædia Britannica* (1947), p. 683 (based on computations by E. Kossina), and are of considerable interest to geographers :—

Name.	Depth in metres (mean).	Area in sq. km.	Volume in cu. km.
Atlantic Ocean	3,926	82,441,500	323,613,000
Indian Ocean	3,963	73,442,700	291,030,000
Pacific Ocean	4,282	165,246,200	707,555,000
I. OCEANS	4,117	321,130,400	1,322,198,000
Arctic Sea	1,205	14,090,100	16,980,000
Malay Sea	1,212	8,143,100	9,873,000
Central American	2,216	4,319,500	9,573,000
Mediterranean	1,429	2,965,900	4,238,000
Intercontinental seas	1,378	29,518,600	40,664,000
Baltic Sea	55	422,300	23,000
Hudson Bay	128	1,232,300	158,000
Red Sea	491	427,900	215,000
Persian Gulf	25	238,800	6,000
Smaller enclosed seas	172	2,331,300	402,000
II. ENCLOSED SEAS	1,289	31,849,900	41,066,000

Name.	Depth in metres (mean).	Area in sq. km.	Volume in cu. km.
Bering Sea	1,437	2,268,200	3,259,000
Okhotsk Sea	838	1,527,600	1,279,000
Japan Sea	1,350	1,007,700	1,361,000
East China Sea	188	1,249,200	235,000
Andaman Sea	870	797,600	694,000
Californian Gulf	813	162,200	132,000
North Sea	94	575,300	54,000
Irish Sea and English Channel	58	178,500	10,000
Laurentian Sea	127	237,800	30,000
Bass Sea	70	74,800	5,000
III. FRINGING SEAS	874	8,078,900	7,059,000
IV. TOTALS OF II AND III	1,205	39,928,800	48,125,000
V. TOTAL FOR ALL OCEANS AND SEAS	3,795 (mean)	361,059,200	1,370,323,000

In their treatise on *The Oceans : Their Physics, Chemistry, and Biology* (1946), Messrs. H. U. Sverdrup, M. W. Johnson, and R. H. Fleming give the same values, in square kilometres, for the oceans and seas as has been quoted in the previous paragraph. Their total for the earth's surface is, however, 510,100,934 square kilometres, and its equatorial and polar radii are given as 6,373,388 and 6,356,912 kilometres respectively. In the following table the area of each 5 degrees of latitude from the equator to the pole is given as percentages :—

Latitudes. (Equator)	Percentage of Hemisphere.	Total Percentage.	Climatic Zones.
0° to 5°	8·68	8·68	Tropics or Torrid Zone
5° ,, 10°	8·62	17·30	,, ,,
10° ,, 15°	8·48	25·78	,, ,,
15° ,, 20°	8·30	34·08	,, ,,
20° ,, 25°	8·04	41·12	23° 30′ (Cancer) Temperate Zone
25° ,, 30°	7·72	49·84	,, ,,
30° ,, 35°	7·36	57·20	,, ,,
35° ,, 40°	6·96	64·12	,, ,,
40° ,, 45°	6·44	70·56	,, ,,
45° ,, 50°	5·92	76·48	,, ,,
40° ,, 55°	5·33	81·81	,, ,,
55° ,, 60°	4·71	86·52	,, ,,
60° ,, 65°	4·05	90·57	,, ,,
65° ,, 70°	3·36	93·93	66° 30′ (Arctic) Polar or Frigid Zone
70° ,, 75°	2·64	96·57	,, ,,
75° ,, 80°	1·90	98·47	,, ,,
80° ,, 85°	1·15	99·62	,, ,,
85° ,, 90° (N. Pole)	0·38	100·00	,, ,,

The North Polar area, or Frigid Zone, includes an area about 8·42 per cent of the Northern Hemisphere (i.e. north of the Arctic Circle at 66° 30′ north latitude. The Temperate Zone from the Arctic Circle to the Tropic of Cancer at 23° 30′ north latitude covers an area 53·87 per cent of the Northern Hemisphere. The Tropics, or Torrid Zone, north of the equator to the Tropic of Cancer, 23° 30′, covers an area of 37·71 per cent of the Northern Hemisphere. These percentages will be on a basis of 98,500,000 square miles for each hemisphere. Therefore, the frigid zones are 8,293,700 square miles each; the temperate zones are also of 53,061,950 square miles each, and the northern half of the Torrid Zone 37,144,350 square miles.

Normal sea water carries about 35 parts of dissolved solids in every 1,000 parts of water, and this is a salinity of 35/1,000, but it may be expressed in percentages, 3·5 per cent, or in parts per million, thus 35,000/per million. Analyses have already been given of the water of the Atlantic Ocean and the Dead Sea (p. 31). However, the following analyses are of some interest as showing the high degree of uniformity of sea water from various seas:—

	(1.)	(2.)	(3.)	(4.)	(5.)
Chlorine	55·24	55·04	55·46	55·53	55·60
Bromine	0·17	0·19	w.Cl	0·18	0·13
Sulphate (SO$_4$)	7·54	7·86	7·59	7·74	7·65
Carbonate (CO$_3$)	0·34	0·18	0·30	0·19	0·02
Sodium (Na)	30·80	30·71	30·53	30·37	30·81
Potassium (K)	1·10	1·06	1·12	1·09	·097
Calcium (Ca)	1·22	1·27	1·21	1·26	0·89
Magnesium (Mg)	3·59	3·69	3·79	3·64	3·74
SiO$_2$, Fe$_2$O$_3$, etc.	—	—	—	—	0·06
	100·00	100·00	100·00	100·00	100·00
Salinity per cent	3·549	3·242	3·465	3·897	3·976

(1) Sample from the Gulf of Mexico.
(2) Sample from English Channel near Dieppe, France.
(3) Sample from North Atlantic between Norway and Iceland.
(4) Sample from Mediterranean, near Carthage (Tunis).
(5) Sample from middle of the Red Sea.

An analysis of normal or average sea water is given below to show the percentage of water and also the more frequent solids which appear to occur as such in solution:—

Percentage of water	96·495 in 100
Sodium chloride	2·700 ,,
Magnesium chloride	0·360 ,,
Potassium chloride	0·070 ,,
Magnesium sulphate	0·230 ,,
Calcium sulphate	0·140 ,,
Calcium carbonate	0·003 ,,
Magnesium bromide	0·002 ,,

Comparisons of average sea water and river water show very considerable differences, as may be noted in the dissolved matter, reckoned in percentages only and not as salinity:—

SALTS FROM OCEANIC AND RIVER WATERS.

	Sea Water.	River Water.	River Water. Less Cyclic Salts per cent.
CO$_3$	0·41 (HCO$_3$)	35·15	35·13
SO$_4$	7·68	12·14	11·35
Cl	55·04	5·68	0
NO$_3$	—	0·90	0·90
Ca +	1·15	20·39	20·27
Mg +	3·69	3·41	3·03
Na +	30·62	5·79	2·63
K +	1·10	2·12	2·02
FeAl oxides	—	2·75	2·75
SiO$_2$	—	11·67	11·67
BrSr, etc.	0·31	—	—
	100·00	100·00	100·00

The change is from water carrying calcium carbonate and dissolved silica to one with sodium chloride as the main constituent. It is thought that the calcium carbonate and the silica is precipitated in sea water by normal processes or by minute animal and plant life in the sea (*Globigerina* and *Coccoliths* respectively for calcium carbonate, and *Radiolaria* and *Diatoms* respectively for silica). The accumulation of sodium chloride, particularly the chlorine, is still insufficiently explained. As F. W. Clarke has said : " In chemical character, fresh and salt water are opposites. . . . We can understand the accumulation of sodium in the ocean, and some of the losses are accounted for, but the great excess of chlorine in sea water is not easily explained."

According to *The Data of Geochemistry* (1924), p. 138, the rivers carry into the ocean 250×10^6 tons of sodium and 150×10^6 tons of chlorine annually, and F. W. Clarke estimated the sodium in the oceans as $12,500 \times 10^{12}$ tons and the chlorine as about $25,000 \times 10^{12}$ tons. On this basis the sea has become salt (as it is) in 50,000,000 years on a reckoning of the sodium, but in 160,000,000 years on a reckoning of the chlorine accumulation. This figure, on the chlorine basis, represents the chlorine in the oceans now, but omits all the cyclic salt (which is carried inland by the winds and redissolved and taken back to the oceans), nor does it include the vast quantities of salt in rock salt deposits in the strata or as salt water held in the strata. However, it is not thought that the total salt which has been carried into the oceans and has (i) remained there, (ii) been cyclic, or (iii) is in rock salt and salt water in the strata can alter the estimate to much more than 200,000,000 years. Since there are extensive deposits of rock salt of Cambrian age (500,000,000 years old) it would seem that there are factors which have not been allowed for in such direct calculations. Indeed, there is no reason to doubt that the oceans as great basins of salt water were already present in pre-Cambrian times. However, J. Joly's original estimate of the age of the ocean (on a sodium chloride basis) was only 90,000,000 years or thereabouts, which few geologists could accept as even approximating to what their field observations and laboratory studies indicate. T. H. Holland and W. A. K. Christie showed that the " cyclic salt " carried by the wind into Rajputana explains the salt of Sambhar Lake, but such a case of excessive cyclic salt may be exceptional. Similarly, we have no real guide as to the total rock salt resources of the world, nor are we satisfied that the oceanic waters always carried the same proportion of salts as now. It has been argued that the excess of chlorine might be due to volcanic eruptions (as suggested by E. Suess (*Geographical Journal*, vol. 20, 1902, p. 520), or from meteoric iron (such as anhydrous ferrous chloride in lawrenceite, or from both these sources (volcanoes and meteorites). Lawrenceite has been detected in the terrestrial native iron of Ovifak.

Sea water dissolves more air in cold water than in warm, and this explains why there is more air in the Arctic Ocean than in the tropical belts in the Pacific, Indian, and Atlantic oceans. Oxygen is, however, more soluble in fresh than in salt water (in the proportion 10 to 9 roughly at the same temperatures). The proportions of nitrogen to oxygen dissolved in fresh water at different temperatures vary from 19 : 10 at 0° C. (volumes) to 14 : 7 at 14° C. (the percentage of oxygen remaining about 33·5—decreasing slightly with rise of temperature). In the case of sea water the proportions are 15·5 : 8·1 at 0° C. (volumes) to 11·3 : 5·8 at 15° C. and 8·35 : 4·15 at 35° (nitrogen to oxygen). The percentage of oxygen falls from 34·4 at 0° C. to 33·3 at 35° C. The pressures are assumed at 760 millimetres of mercury (barometer) and the volumes reckoned in cubic centimetres per litre of water. The dissolved oxygen is thus of great importance, first in giving sustenance to oceanic and aquatic life, and second in oxidizing dead organic matter with the liberation of the ubiquitous carbon dioxide (which might combine with the lime (CaO) and deposit limestone ($CaCO_3$) as carbonate of lime). It is estimated that the oceanic waters contain more carbonic acid (CO_3) in tropical than in cold latitudes. It is to be remembered that variations in the carbon dioxide content of the atmosphere have been considered as a prime factor in the climatic changes which the earth's surface has been subjected to in past geological epochs. T. C. Chamberlin (see *Journal of Geology*, vol. 5, 1897, p. 653, and later volumes) sought to show that carbon dioxide was drawn from its oceanic storehouse. However ingenious Chamberlin's theory may be, it is felt that water vapour is the essential component in regulating the atmospheric temperatures, and hence climate.

L. Brontman has described the observations of the Soviet (Russian) expedition as they floated on the icefloes across the North Pole. He describes the discovery of a warm layer of water in the centre of the Arctic Ocean, showing temperature readings of − 0·48° C. at a depth of 1,500 feet, and even at a depth of 3,000 feet the water was still warmer (− 0·17° C.) than what it should normally have been (− 1·6 to − 2·0° C.). It was, of course, due to the Gulf Stream which had travelled from the Tropic of Cancer across the Atlantic through the Barents Sea to the North Pole. In his account, *On the Top of the World* (1938), Brontman has described (pp. 172 to 177) the finding of " . . . all sorts of sea creatures, some of them microscopic, others visible to the naked eye . . . " when the plankton nets were lowered into the deep and drawn up. He mentions a crab 2 inches long, a bird also which had approached to pick up scraps. The deeper waters could not be measured as the thermometers were unable to stand up to the pressures and had broken. He mentions that the plankton nets when hauled up from a depth of 3,250 feet were literally crawling with all sorts of molluscs, meduras, crabs, etc., with bright red colour characteristic of deep (lightless) sea waters. It was July, and the warm weather thawed the snow and the ice-floes were covered with pools of fresh water which saved fuel and simplified their cooking.

The information given in the previous paragraphs on the extent and characteristics of the oceanic waters has shown the oceans to be a great reservoir as well as vital to the life and climate of the earth. Some idea of the volume of the water stored in the oceans and seas may be gained by working out the volumes given on p. 25, but the total volume of the oceanic waters is of the order of 328,900,000 cubic miles for a superficial extent of, say, 139,400,000 square miles and a mean depth of 12,450 feet. If we take round numbers of 140,000,000 square miles and a mean depth of 12,500 feet, the volume is 331,000,000 cubic miles. This, however, does not take enough notice of the extensive " continental shelves ", and so for simplicity of calculation it is suggested that a depth of 2·25 miles may be as satisfactory as any of the above. This would give the volume of water in the oceans as 315,000,000 cubic miles. If the continental lands were levelled down to a uniform sea floor the mean depth of a completely covering ocean would be roughly 9,000 feet, and its surface would stand at a height of about 600 feet above our present sea-level. The salinity would remain as it is to-day, but truly uniform, as the ocean currents would mix completely the waters throughout this " Flooded Earth " and no land would show anywhere. There would be no mountain chains to precipitate rain or snow, and it is probable that a uniform climate would prevail over the face of this earth ocean. The study of the rocks of the earth's crust, while providing abundant evidence of action by water, both erosion and deposition, does not permit of any deduction in favour of a belief in a completely flooded earth. It is not possible to say with any degree of certainty that the volume of water in the oceans has markedly increased since, say, Cambrian times. Many believe that the oceans have increased in volume and there have been statements to the effect that the level of the Pacific is rising 1 inch every fifty years and that of the Atlantic twice as fast. There are too many factors—wind, movement of uplift or sinking, errors of levelling, etc., which must be fully corrected before any reliable figure can be accepted.

Lakes and Rivers

Although the oceans and seas are the great reservoirs for water on the earth's surface and are supplied with the flow from rivers, there are extensive basins on the land which hold lakes of great size, some of which have no outlet to the sea. The Great Lakes of North America are typical cases of freshwater lakes which overflow to the sea and contain some of the purest fresh water of the world. Other examples are the Nyanzas—Victoria, Albert, Edward, etc., which are drained by the Nile and are also freshwater lakes. Among the closed basin type of lake perhaps the best known examples are (i) the Caspian, which is believed to be a remnant of a widespread sea (the Mediterranean?), (ii) the Dead Sea, which is regarded as always having been isolated but with perhaps an original outflow southwards to the Gulf of Akaba, and (iii) the Great Salt Lake of Utah, which is a shrunken residue of an extensive lake, Bonneville (which, with Lake Lahontan, now also shrunk to small lakes, was a great basin in the western United States). There never seems to have been a seaward outlet from the Bonneville basin, at least since the Quaternary (Pleistocene) Ice Age. Thus these three examples have original differences of some importance, and Utah Lake has suffered change at the hands of man also. Analyses of the waters of these lakes are shown in the following records :—

	(1.)	(2.)	(3.)	(4.)	(5.)	(6.)	(7.)
Cl	41·78	50·26	35·40	63·40	55·48	4·04	26·87
Br	0·05	0·08	—	1·69	—	—	—
SO_4	23·78	15·57	31·27	0·42	6·68	42·68	30·14
CO_3	0·93	0·13	0·10	trace	0·09	19·88	8·48
Na	24·49	25·51	22·05	13·49	33·17	5·81	18·34
K	0·06	0·81	1·07	3·24	1·66	w.Na.	1·75
Ca	2·60	0·57	4·48	5·74	0·16	18·24	5·34
Mg	5·77	7·07	5·40	12·02	2·76	6·08	6·85
SiO_2, Fe_2O_3, etc.	—	—	—	trace	—	3·27	2·23
	100·00	100·00	100·00	100·00	100·00	100·00	100·00
Salinity in parts per 1,000	12·67	163·96	10·67	238·90	203·49	0·306	1·254

(1) Caspian Sea, A. Lebedintzeff, 1905.
(2) Karabogghaz Gulf, Caspian Sea, A. Lebedintzeff, 1905.
(3) Sea of Aral, Stepanow, 1910.
(4) Dead Sea, A. Friedmann, 1912.
(5) Great Salt Lake, Utah, R. K. Bailey, 1913.
(6) Utah Lake, analysis of 1884, by F. W. Clarke.
(7) Utah Lake, analysis of 1904 by B. E. Brown.

All the analyses of the previous paragraph are quoted from *The Data of Geochemistry*, to which the reader is referred for further analyses of these inland seas. I have given analyses of the Sea of Aral because it was under the same sea as the Caspian. The analyses of the Gulf of Karaboghaz is also to illustrate how the Caspian sea water may become concentrated behind a bar. The waters of the Great Salt Lake of Utah has varied in salinity, being as low as 137·90 (per 1,000) in 1877 and up to 277·2 in 1904, but it is nevertheless four to eight times that of normal oceanic water. It was estimated to contain 400,000,000 tons of sodium chloride and 30,000,000 tons of sodium sulphate in 1890. However, it is a different type of water to that of the Dead Sea although the salinity is similar. The two analyses of Utah Lake, which is a freshwater lake, show a great change in twenty years. F. W. Clarke ascribes the increased salinity (four times greater) and the change from a pure water type to a sulphate water as probably due to irrigation. He says : " Its natural supplies of water have been diverted into irrigation ditches, and at the same time salts have been leached from the soil and washed into the lake." The Jordan river drains from Utah Lake into the Great Salt Lake, and an analysis taken near Salt Lake City is shown below with analyses from the Bear river, in Wyoming (Evanston), and again as it approaches the Great Salt Lake from the north (at Corinne). These are of interest as the Bear river also shows a great change in its salinity as well as in its character after its long hairpin loop round the Bear River range in Idaho.

	Bear River (Evanston).	Bear River (Corinne).	Jordan River (Salt Lake City).
Cl.	2·68	32·36	34·76
SO_4	5·76	8·16	30·68
CO_3	52·68	21·53	trace
Na	4·49	20·54	23·04
K with Na	—	—	—
Ca	23·69	10·12	10·26
Mg	6·86	4·76	1·26
SiO_2	3·84	—	—
Al, Fe oxides	—	2·53	—
	100·00	100·00	100·00
Salinity in parts per 1,000	0·185	0·637	1·09

This shows them to be fairly pure waters, except the Jordan river. This like the water of Utah Lake, is of the sulphate type, while that, of the Bear River flowing into the Great Salt Lake is a carbonate water, but both should be tasteless and good drinking water.

Although the Great Salt Lake, now varying from 2,000 to 1,500 square miles or less according to the season of the year, is but a remnant of the Bonneville basin, estimated at 20,000 square miles, and is itself perhaps a portion of a greater freshwater lake in the Great Basin of the western United States, there are existing lakes of greater extent. Among the great freshwater lakes of the earth are :—

	Sq. Miles.	Depth (mean). Feet.	Cu. Miles.
Lake Superior	31,500	500	3,000
Victoria Nyanza	26,000	150	650
Lake Huron	23,000	300	1,400
Lake Michigan	22,400	400	1,800
Lake Baikal	12,750	2,500	6,000
Lake Tanganyika	12,700	2,000	5,000
Lake Great Bear	11,600	270	700
Lake Nyasa	11,000	1,500	3,200
Great Slave	11,000	200	500
Lake Erie	9,940	100	190
Winnipeg	8,500	40	70
Ontario	7,540	400	575
Balkash	7,200	20	30
Ladoga	7,000	300	450

The total water-spread of the above fourteen principal freshwater lakes of the earth's surface approximates to 202,130 square miles, with a

Photo by C. S. Fox.

PLATE VI.—A RADIOACTIVE HOT SPRING IN ABYSSINIA.

The Schmidt Electroscope was used in the field to determine the radioactivity of the water at the spring between Diredawa and Hawash in 1934.

Photo kindly supplied by U.S.A. Information Service, London.

PLATE VII.—MAMMOTH HOT SPRINGS OF WYOMING.

The deposits of calcareous tufa, travertine, build up the walls of the lakes at these springs in the Yellowstone National Park.

[*Facing page 32*

Photo, Wisherd. The National Geographic Society.

PLATE VIII.—OLD FAITHFUL GEYSER in Yellowstone Park.

[*Facing page 33*

cubical capacity of 22,565 cubic miles. It is difficult to estimate the spread and quantity of the water standing in all the other numerous small freshwater lakes and marshes, but perhaps a safe total of all the freshwater lakes may be taken as 300,000 square miles and 30,000 cubic miles (with a mean depth of one-tenth of a mile or 528 feet).

In addition to the freshwater lakes, allowance must be made for salt lakes, like that of Utah, and inland seas, such as that of Aral (30,000 square miles, average depth of 50 feet, and a volume of 285 cubic miles), to say nothing of the Caspian, which covers 170,000 square miles, has a mean depth of 575 feet, and contains roughly 19,000 cubic miles of water. With these would be included the Dead Sea (360 square miles, mean depth 1,000 feet, and volume of 70 cubic miles), Lake Van (200 square miles), Great Salt Lake (2,000 square miles), Lake Chad (Africa), and Lake Eyre (Australia), and numerous

Name of River.	Length in Miles.	Drainage Area. Sq. Miles.	Discharge in Cu. Miles.
Missouri-Mississippi	4,200	1,240,000	155 (annually)
Amazon	4,000	2,722,000	1,000 ,,
Nile-Kagara	4,000	1,107,277	30 (at Aswan)
Congo	3,000	1,425,000	257 (annually)
Lena	3,000	1,000,000	64 ,,
Amur-Kerulen	3,000	785,000	50 ,,
Yangtsi Kiang	3,000	650,000	165 ,,
Yenesei	2,700	790,000	50 ,,
Niger	2,600	580,000	60 ,,
Hwang Ho	2,500	500,000	120 ,,
Obi or Ob	2,400	1,150,000	60 ,,
Volga	2,300	563,000	40 ,,
La Plata	2,300	1,198,000	100 ,,
Mackenzie	2,300	680,000	50 ,,
St. Lawrence	2,100	340,000	25 ,,
Brahmaputra	1,800	360,000	50 ,,
Danube	1,750	320,000	47 ,,
Ganges	1,500	432,000	67 ,,
Indus	1,500	372,000	50 ,,
Orinoco	1,500	350,000	40 ,,

others which may have an extensive spread of water in some seasons during certain years, but appear to be increasing in salinity or slowly drying up. To make a full allowance for all lakes, freshwater and salt or alkaline, is even more difficult than estimating the freshwater lakes alone. However, for general purposes, the water spread of all standing water on the earth's surface might be taken as 500,000 square miles, with a mean depth of one-tenth of a mile and a volume of 50,000 cubic miles. In other words, the freshwater proportion is three-fifths of the whole both for area and volume (with a mean depth in all cases of 528 feet).

When attention is given to the rivers of the world, whether discharging into the oceans (like the Amazon and Mississippi, etc.) or into inland seas (like the Volga and Oxus or Amu Darya) and

Salt Lakes (like the Bear River and Jordan rivers), there are two problems to be settled. One is the volume of water passed along the river annually to its estuary or mouth, the second is the water that may be estimated as a more or less constant volume in the river throughout the year. It is this latter that is the "fixed" volume, while the former is the "current" quantity. Both amounts are not easily determined because of lack of data on the gauging of rivers and the discharge at various times of the year—the Brahmaputra during the rains and winter, the Volga between winter and summer, the Indus in summer and winter, etc. An attempt is made to make some estimates from the data shown on page 31 (collected from various sources).

The above great rivers (twenty) have a total length of 53,650 miles (exclusive of tributaries, except in the cases of the Mississippi, the St. Lawrence from St. Louis, and the Nile with the Kagara). They drain no less than 17,054,277 square miles and convey 2,540 cubic miles of water annually, almost all to the sea, except the Volga. It is probably an under-estimate to reckon the discharge of all the rivers in all countries as conveying 5,000 cubic miles of water to the oceans annually as their "current" account. With regard to the water which remains in the stream beds all the year round, it is perhaps permissible to consider the above twenty rivers as having an average depth of 8 feet and an average width of 480 feet throughout their length of 54,000 miles. This works out to somewhat less than 6 cubic miles, or say nearly 800,000,000,000 cubic feet of water. Including all other rivers and tributary streams, it would seem doubtful if the total can exceed 50 cubic miles as the "fixed" deposit of water in the river beds of the world. This would appear to be a small amount of water. If it was spread uniformly over the earth's land surface the water from the river beds would form a film about 0·0546 in. thick. If the water in the lakes also was spread over the land surface it would cover the ground to a depth of 55·60 inches. The amount from the freshwater lakes would cover the land surface with a layer 33·40 inches deep.

Polar Ice and Glaciers

When dealing with the distribution of water as ice and snow there is again the problem of annual snowfalls and of annual thawing, both of ice and snow. The great storehouse of ice is the Antarctic continent and the lands within the Arctic Circle. Further, large quantities of ice and snow are included in the glaciers of the higher mountains and in the ice-floes of the Arctic Ocean. The north and south Frigid Zones of the earth (within the Arctic and Antarctic circles) cover over 16,600,000 square miles, but much of this in either case is oceanic waters. In neither case is the area entirely covered by ice or snow, and judging by Greenland, the ice-covered continental area in the north, 700,000 square miles, and the Antarctic continent in the south, roughly 5,600,000 square miles of continental ice sheets, we may take polar ice caps as covering an average of 8,000,000 square miles in both hemispheres. To this must be added the area covered by glaciers. The superficial extent of the mountain glaciers is small compared with the extent of the continental ice sheets, but many glaciers may be 1,000 feet thick. In making approximate calculations of the quantity of water stored in the ice-fields and glaciers and as snow in the polar regions and in elevated land areas, it would be advisable to reckon in round figures. If the Greenland ice is 1,000 feet thick and that of the Antarctic also averages 1,000 feet in thickness then it is safe to reckon the ice and snow caps as well as the glaciers and snowfields as covering an area of 8,000,000 square miles to an average depth of 528 feet. This works out to about 800,000 cubic miles.

Compared with the 50 cubic miles of water stored in the river valleys and the 50,000 cubic miles of water in the lakes of the land areas, this 800,000 cubic miles of water stored in the Polar regions and in the glaciers and snowfields of the mountains and high plateaux, is very large. It compares with the oceanic waters, which were calculated at upwards of 300,000,000 cubic miles. If this mass of ice and snow was melted and added to the ocean it would raise its level 30·1 feet. It would cover the land surface to a depth of nearly 74·1 feet and form a layer 21·5 feet deep over the entire surface of the earth. It will be seen from these rough approximations that whereas the water in river valleys is small, and that in lakes much larger, the water held in the form of ice and snow is very large. They represent 50, 50,000, and 800,000 cubic miles respectively. And it may be said that while water in the river valleys (the *fixed*, as distinct from the *current* or flow) is negligible, that in the lakes may affect climate, while that in the ice caps can cause considerable differences in the depth of the water in the oceans. During an ice age, such as the Pleistocene or Great Ice Age of the northern hemisphere (which has almost passed away), the ice and snow might easily have been four times as great as the amount now estimated. The weight would depress the land on which it lay, but the amount removed from the oceans would lower the level of the oceans very markedly. In this manner the Polar ice caps have controlled changes in the distribution of the land and sea and thereby affected the climates of the continental regions.

Water in the Soil

While the alluvium of some valleys and plains may be upwards of 20 feet thick, it has been customary to reckon the thickness of the soil as about the depth to which a plough turns up the surface of the ground. This may be taken as 9 inches on an average, and, allowing for various agricultural considerations, a depth of 3 feet would be a normal thickness of good soil cover. The porosity of such soft material might be as high as 33 per cent on an average, so this amount of soil can hold moisture (water). Allowing for rock outcrops, icefields, and the sandy wastes of desert regions, it is estimated that perhaps one-fifth of the land area of the earth has a thickness of 3 feet of soil holding, say, 1 foot of water. This would mean a little more than 11,500,000 square miles of surface or 2,178 cubic miles. With a pore space volume of 20 per cent the quantity would be 1,320 cubic miles. This might be taken as a rough approximation of the quantity of water which might be held by the soil. It represents 0·12 ft. of water on the surface of the land (57,000,000 square miles). This is roughly 1·5 (actually 1·44) inches of water as a layer on the land surface. It is considerably more than the visible water in the rivers (a layer 0·054 in. on the land), and very much less than that in the lakes (55·6 inches as a layer on the land surface).

Included with the subject of the soil there is the question of the water held by plants and animals on the land. As the moisture in the soil, estimated as shown in the previous paragraph, exceeds that held in the atmosphere, and has been calculated somewhat generously, it is not possible to say what is the excess. In extensive forest areas the water held by the vegetable material might be greater than the moisture in the soil under the trees, but where the soil is not under irrigation and is pasture rather than arable land, it is likely that the average moisture of the soil is less than the general average, and vice versa for irrigated lands. For those who desire some rough approximation of quantities, it may be assumed that the water held as moisture in the soil and in plants and animals as an average is perhaps in the proportion of five to one, so that if the estimate for the soil moisture is a layer of 1·5 inches deep, that for the water held by plants and animals may be taken as 0·3 in. deep. Both estimates are believed to be excessive and may be as little as one-tenth of the amounts calculated, i.e. only 0·15 in. of water in the soil and 0·03 in. for plants and animals, as the thickness of the film a layer of water from these sources would cover on the land surface only.

Water held in the Rocks

There are four categories in which the underground water resources may be dealt with : (i) the water held in combination in capillary spaces or by hydrated mineral substances in the zone of weathering ; (ii) water in movement through the pores and fissures of the rocks to a depth of 2,500 feet ; (iii) water held both as (i) and (ii) from a depth of 2,500 feet to a depth of 12,500 feet, and (iv) water held in

THE DISTRIBUTION OF WATER

combination in the rocks from a depth of 12,500 feet to that of 63,360 feet where the temperature and pressure cause the rocks to be almost at their fusion point. Of these separate types of underground water, (i) is too tightly held to be available for water supply; (ii) is the main source of supply to wells, artesian water, and to springs; (iii) may be partly available from deep borings to below sea-level, and (iv) is with the deeper parts of (iii) the source of hot springs and geysers, as also the zone which may yield the lava for volcanic eruptions, when vast quantities of steam are given off. It is computed that there is a temperature increase of $1°$ C. for every 100 feet of descent in the rocks. If this is uniform the temperature at a depth of 2,500 feet from the ground surface (average height of land above sea level taken as 2,500 feet) is $25°$ C. plus the surface temperature, say, $15°$ C. in the temperate zone, or $40°$ C. ($104°$ F.); at a depth of 12,500 feet the temperature will be $140°$ C. ($282°$ F.), and at 63,360 feet (10 miles) roughly $650°$ C. ($1202°$ F.).

(i) Materials such as kaolin, laterite, clays, etc., may hold from $4 \cdot 0$ to $24 \cdot 0$ per cent of combined water, as is the case with some decomposed rocks, including aluminous laterite which may average 30 per cent of combined water by weight. These rocks are largely at or within a depth of 60 feet from the surface. They are most commonly met with in tropical countries where the land is about 25 per cent of the 74,000,000 square miles, or, say, 18,500,000 square miles. But the area occupied by laterite and related decomposed rocks which hold large amounts of combined water is a fraction of the 18,500,000 square miles. Even if it was $\frac{1}{85}$th part and the rocks averaged only 8 per cent of combined water to a depth of 60 feet, the quantity would not exceed 100 cubic miles, which is twice that estimated for the water in the river beds throughout the world.

(ii) With regard to the free or percolating water in fissures and through the pore spaces of the rocks from the surface to a depth of 2,500 feet (roughly sea-level for broad calculations), the increase of temperature is not to boiling point, and the porosity of the rocks is normally accepted as 4 per cent by volume. Since the land area is roughly 57,000,000 square miles and the depth is roughly less than half a mile, the volume of water is calculated as 1,084,000 cubic miles—a figure approaching that of the water held in the ice and snow of the Polar caps and high mountain regions. This great volume of available water if spread over the land surface would drown it to a depth of $100 \cdot 00$ feet. If added to the oceans it would raise their level (at the existing area) to a height of $40 \cdot 76$ feet. If it was spread over the full extent of the earth's surface it would form a layer of water over $28 \cdot 93$ feet deep. It is this water, below ground, which corresponds to that stored in the lakes above ground.

(iii) The water in the rocks below ground, from sea level to a depth of 12,500 feet (the average floor of the oceans), is taken as partly combined or held very strongly and partly available from the pore spaces of the rocks. The temperature at the level of −12,500 feet, or really 15,000 feet below the surface of the land, is 150° plus 15°, or 165° C. (327° F.), which is well above boiling point at atmospheric pressure. Nevertheless, it is believed that the rocks are strong enough to support the superincumbent weight and therefore for fissures to remain open for the percolation of water. Owing to the great pressure on porous strata, which still permit water to percolate, the pore space volume is reckoned at only 1·0 per cent. Excluding complications of " continental shelves " and assuming the surface area at the same total of 57,000,000 square miles, the volume of water stored in this zone of the rocks works out to 1,349,100 cubic miles. This quantity would cover the land surface to a depth of 125 feet. It would raise the existing ocean surface 50·17 feet, and if spread over the entire surface of the earth would drown it to a depth of 35·60 feet.

(iv) As regards the water held in the subcrustal zone to a depth of 10 miles below the −12,500 feet of the ocean floor (as averaged), assuming for convenience a temperature of 0° C. at the ocean floor, the temperature at a depth of 528,00 feet should be 528° C. (986° F.) below the oceans, and about 678° C. (1436° F.) below the land surface (67,800 feet above). Assuming a pore space of 1·0 per cent in the rocks, which are heated and with little possibility of percolation, the water held in combination or very tightly in the pore spaces amounts to 19,700,000 cubic miles in this subcrustal zone girdling the subcrust in a stratum 52,800 feet thick. If so large a volume were added to the surface of the earth it would cover it to a depth of 528 feet. From these calculations, on the assumptions made, the total amount of water held by the rocks of the earth's crust, under categories (i), (ii), (iii), and (iv), the total volume would be barely one-fourteenth of that in the oceans.

(v) Perhaps the depth of $12\frac{1}{2}$ miles below sea-level is not sufficient to limit the source of water vapour within the rocks of the earth's crust, and the so-called rock magma at a greater depth may be the greatest source of water make up to losses from the ocean. Geologists, seismologists, and others have expressed opinions of considerable interest on this problem, but we still lack the data that must decide the questions at issue. Appreciable quantities of water have been collected from the molten lava of Kilauea, and it is well known that very large volumes of steam (far greater than all the other volcanic gases together) are emitted during eruptions. Analyses of the gases taken from Vesuvius show temperatures of 250 to 300° C. and have between 67 and 78 per cent of water vapour (by weight). And there can be no question that the molten lava before its eruption must be like soda water in a siphon bottle. When the siphon is opened the gas emerges with violence and carries the water with it. This is perhaps how volcanic eruptions occur, but it does not prove that the contained water has come from a greater depth than $12\frac{1}{2}$ miles, nor does it give the data for estimating the water in the heated rocks and the magma. However, there is a belief that the make up of the oceanic waters comes from the subcrustal rock magma at considerable depths.

NOTE

It may help to simplify the calculations relating to the surface of the land and oceans of the earth and to the depths of the ocean and the sub-crustal layer if we adopt the following measurements: for 2,500 feet take 2,640 feet or half a mile (which is roughly 0·8 kilometres); for 12,500 feet take 13,200 feet or 2·5 miles (which is about 4·0 kilometres); and for 63,360 feet take 66,000 feet or 12·5 miles (which is nearly 20·0 kilometres). Also reckon 197 million square miles as equivalent to 510 million square kilometres, and also 140 million square miles roughly equal to 362 million square kilometres, and thus 57 million square miles as the same as 148 million square kilometres. And since one cubic mile equals 4·165 cubic kilometres, or one cubic kilometre equals 0·2403 cubic miles, and a cubic mile of water weighs 4,089 million (say 4,090 million) tons, the equivalent is 4,155 million metric tons (or $4·155 \times 10^{12}$ kilograms or $9·15 \times 10^{12}$ pounds).

CHAPTER III.—THE CIRCULATION OF WATER

Although the distribution of the water of the earth has been briefly discussed in the previous chapter, the manner of treatment was on the presumption that the water was " fixed ". Actually, there is continual movement of the earth's water. The main reservoir is the ocean. From its surface water vapour is formed by the heat from the sun and carried up into the atmosphere by the air movement we call winds and breezes. The water vapour in the air is condensed to drops as the air rises and becomes visible as clouds. The contained moisture may then be precipitated as rain or snow or hail or mixtures of these. The rain or snow which falls on the land may be partly re-evaporated, a part may flow into the streams and rivers and be returned to the ocean, and a part may sink into the ground where it supplies the moisture to the soil and also infiltrates downward into the rocks. The infiltrating water may re-emerge as springs, but some of it reacts on the rocks and some goes deeper, either trapped in sedimentary formations or held in the capillary spaces, pores and fissures, in the rocks. Some of the water carried to great depths may be forced out and may return as thermal water ; some may enter into combination with the heated rocks and be given back to the atmosphere and ocean during volcanic eruptions. Thus, there is ceaseless movement and action, both in the ocean and in the atmosphere by the wind, and below ground by the percolating water. This is what Oscar E. Meinzer has called the " hydrolic cycle " (*Physics of the Earth*, Vol. IX : Hydrology).

The Heat from the Sun

Geologists believe that the sun has been radiating heat for at least 1,500,000,000 years, or since the earliest known sediments were deposited by moving water. Physicists trace the age of some rocks (earliest gneisses and schists) back 2,500,000,000 years on the basis of radioactivity and the amounts of helium, lead, or other products formed. There is little doubt that water was in circulation on the earth's surface long before the beginning of the Palæozoic era 500,000,000 years ago, since ripple-marked surfaces and raindrop splashes have been recognized on the bedding of some pre-Cambrian strata. It is not possible to say if the quantity of water in the oceans was larger than at present, but the evidence generally, from the history of the rocks (the *Geological Record*) is that while there have been great changes in the climate of various regions—an Ice Age in the southern hemisphere in Carboniferous times almost at the same epoch of tropical conditions in the northern hemisphere, and subsequently a great Ice Age in the northern hemisphere in Pleistocene times—there is little evidence to support the idea that the earth's surface has been cooling steadily during the past 500,000,000 years. And there is no doubt that without the sun's warmth the earth's surface would be at a freezing temperature.

The sun radiates heat in all directions. The face of the sun which is seen from the earth is the photosphere and is heated to a temperature estimated at 6,000° C. It is incandescent. The main body of the sun is never visible. Although the true volume of the sun is larger than its limit at the photosphere (a diameter of 840,000 miles), as seen during an eclipse of the sun (when corona and other prominences become visible), its distance from the earth varies from 94,500,000 miles in July to 91,300,000 miles in January. At these distances and considering the Earth's diameter is less than 8,000 miles (the polar and equatorial diameters are roughly 7,899·6 and 7,926·6 miles respectively), the earth receives a minute part of the energy radiated from the sun. The amount is estimated at $\frac{1}{2000000000}$th part. The quantity received on the outer atmosphere of the earth has been computed at 1·95 calories per minute per square centimetre. In its passage through the atmosphere to the earth's surface, in the tropics, some of the energy is lost, absorbed in ozonization in the upper air and to the water vapour nearer the surface in the troposphere. The energy at the earth's surface, say, in the Pacific or Indian Ocean, is assumed to be 1·33 calories per minute per square centimetre, but 10 per cent of this appears to be reflected back into the atmosphere. There are thus roughly 1·2 calories per minute per square centimetre available for heating the land or evaporating water on a sunny day in the tropics. In terms of active energy the sun's heat, if converted into power, would be equivalent to 4,500 horse-power or 3,357 kilowatts continuously from each acre on which the sun's rays fell.

Evaporation from the Sea and Land

In all these descriptions the tendency is to forget that although the sun's rays fall steadily on the earth's outer atmosphere and penetrate to the surface of the oceans and continents, the earth itself is a sphere and rotating on its axis. The rotation is nil at the poles, but is at 500 miles an hour at a latitude of 60° and twice that speed at the equator. The axis is tilted to the plane of its orbit round the sun, but the axis is more or less fixed. This means that a body resting on the earth's surface tends to be flung off at a tangent by the rotation while prevented from doing so by the force of gravity. As the earth rotates from west to east there is a tendency for the oceanic waters along the equatorial belt to be left behind or to flow to the west as equatorial currents. This, in fact, they do, but owing to the distribution of land and sea these equatorial currents, especially in the Indian Ocean, develop counter currents as eddies. It is due to the equatorial currents flowing westwards that there is an eastward current (the west wind drift in the open stretch of ocean between south latitudes 40° to 60°, and the northward currents up the west coasts of South America (Peru current), Africa (Benguela current), and Australia and southward currents down the west coast of North America (California current) and Africa (the Canaries current). In the Atlantic the north equatorial current forces itself into the Gulf of Mexico and emerges as the Gulf Stream. Similarly, the south equatorial current meeting the coast of Brazil sweeps southwards and so makes a vast circle with the Benguela current, the west wind drift and the equatorial current.

There is more freedom for the atmosphere to circulate under the spin of the earth's rotation as the oceanic waters tend to do, but the heating effect of the sun's rays causes the atmospheric circulation to be upwards as well as horizontal. At the equator, where westerly winds might be expected, the air is heated and rises. As a consequence, the cooler air from the temperate regions to north and south flows in with a westward component and so causes the south-east trade winds and north-east trade winds of the Pacific and Atlantic oceans. There are, corresponding to the west wind drift of the southern hemisphere, strong westerly (from west) winds in the same latitudes (40° to 60°). All these winds are more or less permanent. Complications occur in the Indian Ocean where the continents of Africa, Asia, and Australia interfere with both the ocean currents and the air currents. The main differences are due to the unequal rate of heating of the land as compared with the sea surfaces. The specific heat of water is five times greater than the rock of the land. In consequence of this unequal heating of the water and land surfaces (and the rapid cooling down of the land surface after sundown) the air in contact with these surfaces is heated and rises more quickly in hot land areas. Where the air rises above warm water surfaces it carries away water vapour. This removal of water from a surface exposed to an air current is called evaporation.

The following table gives an approximate idea of the ability of air to take up moisture as water vapour at different temperatures and the vapour tension that is exercised by the water vapour in the air under normal conditions of atmospheric pressure :—

Temperatures in		Vapour Tension in		Water per cu. ft. of Air in	
°C.	°F.	mm.	in.	grains.	grams.
−10	14	2·16	—	—	—
0	32	4·58	0·10	2·37	0·153
10	50	9·21	0·36	4·28	0·277
21	70	18·65	0·73	8·00	0·52
33	90	36·83	1·45	15·43	1·00
45	113	73·66	2·89	31·00	2·00

From these data it will be seen that air at 90° F. may be carrying upwards of 10 grains of moisture per cubic foot and yet feel dry, while air at 50° F. may be saturated with moisture and yield rain although it contains less than 5 grains of moisture per cubic foot. A cold wind, say, at 32° F., coming from a mountain upland may as it descends and warms (say, to 70° F.) be capable of taking up more than 5 grains of moisture per cubic foot. The Fohn in the Alps and the Chinook of the west (U.S.A.) are cold drying winds of the type mentioned. Indeed, it is on record that the Fohn has evaporated 2 feet of snow in the Grindelwald valley in twelve hours. In the case of the Chinook, at Kipps (Montana), pasture land was covered with $2\frac{1}{2}$ feet of an early snowfall and an ugly situation for the cattle farmers was cleared in less than an hour by a blast of the Chinook which was warmed from 45° ($-13°$ F.) below zero to 24° F.—a rise of 37° in a few minutes.

Dry, cold, clear air absorbs very little heat, and it is not an uncommon experience for climbers to be standing in snow while in an icy air and yet feel the effects of the sun's heat on their faces and head by direct radiation. At normal altitudes the temperature of the air increases 1° F. for every 324 feet of descent, and vice versa. At higher altitudes the amount increases to 540 feet (for 1° F. rise or fall). At great heights (by plane) the change of level may exceed 1,000 feet for a rise or fall of 1° F. Put in terms of the mercury barometer (760 mm. or 29·922 inches at 0° C. or 32° F. at Paris, or 29·02 inches at London), equal to 14·75 lb. per square inch, it means $\frac{1}{10}$ in. fall of the barometer corresponds to a rise of 100 feet; at an elevation of 18,000 feet a similar fall of the barometer indicates a rise of 200 feet; at altitudes of 36,000 feet it means a rise of 400 feet, and at 60,000 feet it means 1,000 feet. It is assumed that the air temperature has remained at 0° C. At lower altitudes, within the troposphere (32,000 feet), an appreciable rise or fall in elevation is accompanied by a fall or rise in temperature, corresponding with an expansion or compression of the air. This explains how the Fohn is warmed as it descends and, becoming drier, is able to evaporate snow or moisture.

All the phenomena of cloud formation—rain, snow, dew, and condensation of water on rock surfaces from moisture or water-vapour-laden air—are based on the principle that air is able to take up definite amounts of water vapour at definite temperatures under atmospheric pressure. If the air pressure falls, expansion and fall of temperature usually follow and, similarly, if the air is compressed, its temperature will rise and increase its power to absorb more water vapour. Modifications of the principle occur, such as in sea and land breezes at Gibraltar. During the day the quickly warmed rock draws the breeze from the sea and causes it to rise along the steep rock face. At night the rock cools more rapidly than the adjacent sea, and an off-shore breeze brings the damp air down against the rock face and thus obtains water for drinking purposes. The unpleasantly warm and humid air in the Persian Gulf would appear to be capable of depositing a copious dew on the high lands of Oman in the Strait of Hormuz (Jabal Al Harim stands above 6,000 feet high), but the country there is a picture of waterless desolation. However, the principles remain unaltered for the evaporation of water by dry warm air, and condensation from moist air if cooled below its saturation or dew point. Various factors affect the process of evaporation, as already shown, but the sun is the source from which the water or land is heated, and by contact with such surfaces dry air is warmed. It is the warm dry air that takes up the water vapour and carries it away to the land.

Another effect that is noticeable when evaporation occurs is that of cooling. Heat is taken from the air to evaporate the water and the temperature of the air falls as a result. In this phenomenon there is involved not only the high specific heat of the water but also the heat of vaporization (latent heat of steam). The character of the climate of any place is dependent on the behaviour of water under different conditions of temperature in that area. If the evaporation of water cools the air the reverse action must warm the air, since the absorbed heat must be given up again. A fall of rain in an arid region often produces a lowering of temperature, but this would appear to be actually due to an appreciable re-evaporation from the originally dry and hot land on which the rain fell from a cloudburst or sudden development of rain cloud in the air above the desert. Since heat is absorbed in the process of evaporation and the air is cooled at the same time, there is a lag in the operation. Similarly, when air is sufficiently cooled to permit condensation of the water-vapour, heat is given up and again there is the tendency to retard the condensation. It is not uncommon in very hot countries for evaporation to continue in the evening, and in consequence for the temperature to fall to freezing point owing to the heat absorbed in vaporizing the water exposed to the cooled air.

It is observed that evaporation decreases with the salinity of water, and thus salt water evaporates somewhat more slowly than fresh water. Other conditions being equal evaporation is slower from large water surfaces than from small sheets of water, but several factors affect the rate at which water particles are taken into the air—the heat supply is essential, but the condition of the air has considerable influence. The water particles may escape into the air by simple diffusion, or by normal convection, or more easily with a gentle breeze. A fall in the atmospheric pressure (barometer) tends to increase the rate of evaporation. Evaporation is naturally more rapid in hot countries and from shallow lakes and reservoirs in such regions. Ice and snow are both subject to direct vaporization (evaporation) during a thaw, but the action may take the form of sublimation pure and simple and the ice or snow merely disappear into the air as water vapour. It is estimated that an evaporation of somewhat less than 0·1 in. may take place in a day in the Atlantic Ocean, with an annual total of 3·8 feet. Other data suggest that the evaporation in the Pacific and Atlantic oceans may be 2·5 feet in 50° north latitude, 5 feet in 20° north latitude, 4 feet in 20° south latitude, and 2 feet in 40° south latitude. In the Great Lakes region of North America the annual evaporation is given as 1·8 feet and from 2 to 6 feet in the land areas of the United States. In an arid and hot region during a period of steady wind the evaporation from a water surface may exceed a rate of ·75 to ·50 in. in twenty-four hours, but amount to over 7 feet as the annual evaporation.

All air in the atmosphere contains water vapour. The average of normal air within an altitude of 7,000 feet is probably 1·4 per cent of water vapour, but about 4·0 per cent has been assumed as normal for the air close to the ground. Much depends on local conditions and on the season of the year in different places. There will be more water vapour in the air in the summer than in the winter, although the air may feel drier in the warmer weather. The humidity is ascertained by an instrument called a hygrometer, which has been designed to read the degree of saturation (relative humidity) of the air. In places in the Sind desert and Rajputana the air may be so dry as to give a negative reading for humidity. Estimates of 1·7 grains of water per cubic foot of air are not uncommon during the dry hot months in Central India, while 11 grains of water per cubic foot of air is not unusual during the rains in Bengal. During the month of May the evaporation in the Bay of Bengal may be from $\frac{1}{3}$ to $\frac{1}{4}$ in. per day, or equal to 0·25 lb. of water per square foot or roughly 580,750 cubic feet from each square mile or roughly 15,000 tons of water. Taking the Bay of Bengal as 1,200 miles across and 1,200 miles down, and allowing evaporation at $\frac{1}{10}$ in. for thirty days, would mean the removal of approximately 68 cubic miles or more than 278,000,000,000 tons of water, all of which would go with the wind into the region at the head of the Bay. Actually, each year the evaporation from the Bay of Bengal cannot be less than 3 feet, or twelve times greater than the above estimate for thirty days.

In endeavouring to secure an average figure for the annual evaporation from the oceans, lakes, icefields, etc., on the earth's surface it is to be remembered that averages are merely for general calculations and of little use for any particular area. Evaporation in any area is affected by (i) difference of vapour pressure (tension); (ii) direction and velocity of the wind; (iii) air temperature by day and by night; (iv) heat supplied for formation of water vapour; (v) heat storage in the air and water; (vi) atmospheric pressure, and (vii) character of the water, whether fresh or salt. So far as the

140,000,000 square miles of the oceanic surfaces are concerned, and taking notice of the evaporation of ice and snow, it may be permissible to suppose an average evaporation of 24 inches over the entire area. For the water spread of lakes and rivers and marshlands an area of 500,000 square miles of surface with an evaporation of 48 inches is assumed. For the icefields and glaciers an area of 16,000,000 square miles is taken with an evaporation of 24 inches. These figures work out roughly to 53,030 cubic miles from the oceans, 378 cubic miles from the lakes and rivers, and 6,048 cubic miles from the icefields and snow, or a total of 59,456 cubic miles, say 60,000 cubic miles of water evaporated by the sun annually from the oceans and land areas of the earth.

No allowance is made for re-evaporation after rainfall, or transpiration by plants and animals. It is admitted that the evaporation from the lakes and rivers may be greater than estimated, but even twice the quantity will make little serious difference to the total.

Precipitation as Rain and Snow

There is always water vapour in the air, less in total amount in cold weather and more and more the warmer the air over a given area. In hot, arid regions, like the Sahara or the Indian desert, the air may be warm but yet low in humidity because there is little moisture to evaporate from the country around. In England, when the weather has been warm, as in summer, the air takes up moisture which, as the air tends to become saturated, generally means rain if not a thunderstorm and heavy rain. The warm air with its water vapour rises and cools, and gives rise to the " wool " clouds which are so evident on sunny days. The upward current under these *cumulus* clouds may be felt by aircraft and have been used by glider pilots for gaining height. Condensation of water vapour occurs as soon as the air temperature is cool enough to produce saturation (about the base of the cumulus cloud) when the fine droplets of water form and make the air (saturated with moisture) visible as clouds. If the temperature of the air is at freezing point, the moisture forms ice crystals which float as flakes of snow. Snowfalls in cold climates are the equivalent of rain in temperate climes. If the prevailing winds are moisture-laden and the land is cold, snowfalls occur frequently, say, around Ben Nevis (Scotland), where there may be fifty days of snow. Around Falmouth (England), where the temperatures are higher, there are barely five days of snowfall a year. The rainfall at Ben Nevis averages 160 inches a year, spread over 264 days or so, whereas the rainfall at Falmouth is barely 45 inches, spread over 205 days.

It does not follow from what has been stated above that either Ben Nevis or Falmouth is famous for fogs or mists and enjoys relatively few sunny days. There are probably only a few hours of dense fog—

perhaps not five days of fog—about Ben Nevis annually, while Falmouth may have five days of dense fog and twelve days of normal fog in a year. The days of good clear weather at Ben Nevis probably exceed 200, and in the case of Falmouth the figure must exceed 275 a year. The corresponding figures for London (Kew) are roughly two days dense fog, thirty-five days fogs, and 110 days clear weather annually. And at Manchester, where a visitor feels as if it were always raining, the figures are eight days dense fog, fifty-five days fog, and eighty clear days. The average annual rainfall, rainy days, and wettest days in a few places in the British Isles is given below :—

	Rainfall, in in inches.	Rainy Days a Year.	Wettest Day, in inches.
Ben Nevis	160	264	4·96
Aberdeen	30	215	1·26
Glasgow	37	200	1·40
Buttermere	105	230	3·85
Manchester	32	195	1·20
Birmingham	28	180	1·35
Nottingham	24	178	1·27
London (Kew)	24	167	1·26
Margate	23	166	1·18
Bournemouth	32	168	1·32
Falmouth	44	205	1·60

From the above figures (taken largely from data compiled by Arnold B. Tinn in his *This Weather of Ours*, 1946), it would seem, statistically, that Manchester and London do not compare unfavourably with Margate and Bournemouth for dull days. As regards hours of sunshine annually, the records show Margate as enjoying 1,780 hours, Bournemouth 1,758 hours, Falmouth 1,710 hours, Tunbridge Wells 1,630 hours, Bath 1,540 hours, Oxford 1,500 hours, London (Kew) 1,470 hours, Birmingham 1,300 hours, and Manchester about 1,000 hours only. The seasonal temperatures in these localities are :—

	January Averages. (max.) Winter. (min.)		July Averages. (max.) Summer. (min.)	
Ben Nevis	27·5	20·6	44·6	37·6
Liverpool	44·7	36·7	65·5	54·5
Nottingham	43·9	34·4	69·1	53·1
London (Kew)	44·9	36·0	71·1	54·9
Margate	44·8	36·7	68·5	56·1
Falmouth	47·7	39·9	66·5	55·2

The coldest and warmest temperatures at these localities have been below and above the averages given, having been to 0° F. on Ben Nevis and as low as 9° F. in London in winter, and as high as 94° F. in London in the summer, and no higher in Margate or elsewhere in the British Isles. Thus, taken " by and large ", the climate in the London district (or lower Thames valley) is probably as good as anywhere in England, except for the fogs which shroud the big cities—Manchester (64 days), Nottingham (85 days), Birmingham (37 days), London (37 days). Falmouth has fifteen days of fog, but it is a sea fog.

Average rainfall is of importance to engineers in all water supply problems, where data over thirty-five to forty years can give an expectation for storage in a reservoir. Minimum rainfall data are valuable as an indication of what may be expected in years of drought so that extra storage capacity may be allowed for. In his most informative book, *The Nation's Water Supply* (1936), which, incidentally, deals with the British Isles only, Mr. R. C. S. Walters has provided maps instead of tables of rainfall. It must be admitted that such coloured maps of average rainfall, minimum rainfall, the minimum that has occurred as a percentage of the local average rainfall, and maximum rainfall are simple and most useful. The areas of greatest annual rainfall are western Scotland, the Lake District, and Wales with spots in other areas, such as Dartmoor. However, excessive rainfall is often a great danger to reservoirs and in causing serious floods either directly or by the collapse of dams and embankments. A rainstorm in Lincolnshire, at Louth, on 29th May, 1920, yielded 4·69 inches in three hours, a flood of 5,000,000 tons of water fell on 22 square miles of country causing damage estimated at £100,000. Another storm, in Norfolk, around Norwich, on 26th August, 1912, yielded 3 inches over 3,463 square miles, with 7·3 inches at Norwich. The record fall for the British Isles appears to be that at Bruton, in Somerset, when 9·56 inches fell on the 28th June, 1917.

It is noteworthy that the heavier rainstorms, or cloudbursts in many cases, are part of great thunderstorms and generally occur in areas of low rainfall and warm to hot regions. In many cases where there is a great display of lightning, accompanied by thunder, there may be very little rain. On 9th to 10th June, 1923, there were nearly 7,000 flashes of lightning over London in six hours when 2·56 inches of rain fell at Hampstead (London). The rainstorm covered the region from Sussex to Lincolnshire, yielding an average of 3 inches over the area, but with 4·04 inches at Newhaven and 4·55 inches at Rottingdean. Another thunderstorm over London, 16th June, 1917, produced 4·65 inches in $2\frac{1}{2}$ hours, almost entirely north of the Thames (4 inches on Campden Hill, 3·65 inches on Kensington Gardens, but none in the area from West Ham to Wimbledon. A storm over Somerset, Cannington, on 18th August, 1924, produced 9·40 inches of rain, of which 8 inches fell in five hours. It has been recorded that 1·25 inches of rain fell in five minutes at Preston, on 10th August, 1893. These figures from the British Isles, remarkable as they are, both for heavy, widespread rain and actual downpours have, of course, been exceeded in other countries.

The following rainfall records covering a period of thirty years shows the very heavy precipitation (which is normal) on the Assam plateau, India, probably one of the heaviest rainfall regions in the world. At Cherrapunji the rain falls largely at night, and since the country is elevated and open, the water drains away rapidly :—

PLATE IX.—THE VALLEY OF TEN THOUSAND SMOKES, ALASKA.

Photo, Kolb, The National Geographic Society.

Photo, Jack Breed. The National Geographic Society.

PLATE X.—NATURAL BRIDGE, GROSVENOR ARCH, UTAH.

THE CIRCULATION OF WATER 49

	Cherrapunji. in.	Maoflong. in.	Shillong. in.	Jawai. in.
January	0·74	0·87	0·49	1·07
February	2·16	0·69	0·81	2·04
March	11·08	1·93	1·85	6·30
April	32·24	4·81	4·29	10·46
May	51·53	11·41	10·06	26·18
June	105·12	32·11	16·46	66·15
July	109·49	30·00	13·48	43·94
August	76·50	21·10	12·79	34·74
September	53·25	19·26	14·75	31·66
October	13·97	8·73	6·23	12·69
November	1·49	0·47	0·98	1·44
December	0·23	0·28	0·25	0·75
Total	457·800	131·66	82·44	237·42

It is to be mentioned that the maximum recorded rainfall at Cherrapunji is 905 inches (1861), of which 366 inches were precipitated in July. The maximum rate of precipitation in twenty-four hours at Cherrapunji was 41 inches on the 14th June, 1876. The greatest recorded rainfall for twenty-four hours was the 46 inches at Baguio, in the Philippines, in June, 1911. A record of 1·02 inches of rain in a minute is claimed for Opid's Camp, in the San Barriel range, California, in April, 1926. It is certain that " cloud bursts " have discharged great volumes of rainwater at faster rates than the highest which have been so far recorded. Many reliable observers have seen the rain coming down literally " in sheets " or " solid ".

In contrast with the above, the following rainfall data from Jaisalmer (Rajputana), in the Indian desert, will be of interest :—

Rainfall in inches,, at Jaisalmer.

	1935.	1936.	1937.	1938.	1939.	1940.	1941.	1942.	1943.	1944.
January	0·40	—	—	—	—	0·55	—	0·34	0·19	1·31
February	—	0·18	1·81	—	0·49	0·27	—	0·75	—	0·84
March	0·17	—	—	—	0·44	0·35	—	—	—	0·13
April	0·22	—	0·05	—	—	—	—	—	0·15	0·42
May	—	—	—	—	—	0·12	—	0·79	—	0·07
June	—	1·36	—	1·46	—	0·22	—	—	0·50	—
July	5·23	2·84	5·91	0·41	—	1·22	3·86	2·05	2·67	7·31
August	2·21	0·22	—	2·24	0·64	3·20	0·23	2·76	0·18	9·80
September	0·61	0·21	1·32	—	0·12	—	0·49	—	0·15	0·73
October	0·10	—	0·05	—	—	—	—	—	—	—
November	—	0·63	—	—	—	—	—	—	—	—
December	—	0·29	0·66	0·15	—	—	—	—	—	—
Totals	6·94	5·73	9·80	4·26	1·69	5·93	4·58	6·69	3·83	20·64

The average rainfall in Jaisalmer over a period of fifty years is 6·5 inches a year. The minimum was the figure of 1939 of 1·69 inches, and the maximum figure appears to have been 20·64 inches in 1944. In 1944 some of the rain in July and August fell in torrential downpours and led to flooding. Normally, the average rainfall allows of some cultivation in the hollows between the sand dunes (which hold water and seepages add to the valleys). When the rainfall is less than 5 inches the people move away with their flocks and herds, but return the following year.

The following rainfall records of the Nile drainage are of some interest. They cover the entire catchment of over a million square miles, from the region of Victoria Nyanza, the White Nile or Bahr el Jebel, the Sobat in Abyssinia, the White Nile in the Sudan, the Blue Nile in Abyssinia, the Nile in the Nubian desert, and the Nile in Egypt. The rainfall is given in millimetres (1 inch equals 25·4 millimetres):—

	(1.)	(2.)	(3.)	(4.)	(5.)	(6.)	(7.)	(8.)
January	150	15	10	10	10	10	15	40
February	150	10	10	10	10	10	15	40
March	200	25	15	10	20	10	10	10
April	250	100	40	15	30	10	10	10
May	200	125	100	25	50	10	10	10
June	50	125	150	150	150	10	10	10
July	40	150	200	100	150	10	10	10
August	50	150	250	125	300	15	10	10
September	75	100	125	50	100	10	10	10
October	125	50	50	25	50	10	10	10
November	150	25	15	10	10	10	10	20
December	150	20	10	10	10	10	15	50
Totals	1,600	895	985	540	990	125	135	230

(1) Rainfall, in millimetres, on the Victoria Lake plateau.
(2) ,, ,, ,, along the valley of the Bahr el Jebel.
(3) ,, ,, ,, of the Sobat basin, Abyssinia.
(4) ,, ,, ,, in the valley of the White Nile, Sudan.
(5) ,, ,, ,, in the valley of the Blue Nile, Abyssinia.
(6) ,, ,, ,, in the valley of the Nile in the Nubian desert.
(7) ,, ,, ,, in the valley of the Nile, Aswan to Cairo.
(8) ,, ,, ,, in the valley of the Nile below Cairo.

It is computed that the average rainfall on the land surface of the earth is roughly as follows:—

```
 6 per cent receives 75 inches or more, averaging 75·0 inches.
16   ,,      ,,     50 to 75 inches        ,,     62·5   ,,
25   ,,      ,,     25 to 50    ,,         ,,     37·5   ,,
33   ,,      ,,     10 to 25    ,,         ,,     16·5   ,,
20   ,,      ,,     under 10    ,,         ,,      5·0   ,,
```

From these figures it would appear that 53 per cent of the land surface averages under 12 inches of rainfall and corresponds to the arid regions of the earth, the Indian desert, the Sahara, etc., and it is necessary to reckon the land area on which rain falls as perhaps 42,000,000 square miles, while the remaining 15,000,000 square miles is the surface on which snow falls and ice accumulates. It is difficult to estimate the amount of snowfall. The average figure is about 10 inches of snow to 1 inch of water, but the snow itself varies from "light", with 0·5 per cent of water content, to *hard* snow with 40 per cent. It is presumed that the snowfall on the 15,000,000 square miles of the earth is equivalent to 24 inches of water or rainfall annually, and that the rainfall on the remaining 42,000,000 square miles of the continents, etc., averages 36 inches of rain a year. The totals of these figures work out to 5,681 cubic miles for the snowfalls and 23,903 cubic miles for the rainfall, both on the continental and other land areas.

It is to be remembered that very large quantities of ice go directly into the oceans as icefields discharge into the sea around Greenland and the Antarctic, also many rivers, such as the Lena, Yenesei, and Ob in Siberia and the Mackenzie in Canada deliver their waters into the Arctic Ocean. Also there is considerable direct rainfall and snowfall on to the ocean surfaces. It is therefore largely an academic calculation as to what proportion of the total evaporation falls on the land areas. Suffice it to say that the oceans supply by evaporation by far the larger portion of the rain and snowfall of the land areas. This is evident when we compare the evaporation from the land to the rainfall and snowfall on the land as shown below :—

Evaporation from—		*Precipitation on—*	
Oceans	53,000 cu. miles	Oceans	29,842 cu. miles
Lakes, etc.	378 ,,	Land as rain	23,903 ,,
Snow and ice	6,048 ,,	Land as snow	5,681 ,,
Total	59,426 ,,	Total	59,426 ,,

Allowing that the evaporation is balanced by the precipitation, the above gives the account in broad outline, but we are not so sure that water vapour is not lost into space, nor the rate at which water is absorbed or/and trapped by the rocks and sediments, nor yet again of the rate at which water is given back or added to the oceans by volcanoes or other subterranean outlets, such as geysers and thermal springs.

Run-off Rainfall

Run-off rainfall refers to that flowing back to the oceans through streams and rivers, and includes glaciers and melting ice which go directly into the sea as icefloes and icebergs. In the main, run-off comes as a result of rainfall, but in some cases it must include the water from springs which are clearly supplied from immediate rainfall. Of the water that falls on the land as rain or snow it may be said that :—

      (i) Some is immediately re-evaporated ;
     (ii) Some seeps into the ground by percolation ;
    (iii) Some is absorbed by the soil and plants ;
    (iv) Some runs off directly into the streams ; and
     (v) Some is given up to the streams by springs.

The proportions vary very considerably according to climate and season and physical conditions, including the geology of the areas considered, and often according to the time of the year and local factors in the same area. A gentle fall of rain in a hot country on to a porous surface would not furnish any rainfall if the rain was not prolonged until the ground was saturated and the air cooled down. A similar amount of rain in a temperate climate on a well drained area might flood the streams since there might be little evaporation and negligible percolation. With the excellent drainage of the country, in the Thames Valley every shower tends to flood the river.

All water supply questions are bound up with geological considerations, and thus the nature of the rocks exposed in a given catchment, as well as their structure or mode of occurrence, affect the run-off rainfall. If the country is hilly and underlaid by slates and similar impervious strata there will be a larger proportion of run-off than from gentle slopes of soft sandstones or other porous beds. In limestone regions, where the strata are bedded and jointed and subject to solvent action by rainwater, practically little run-off may result even from heavy rain. Where elevated land is cut by deep river channels and the strata above river bed level are porous, much of the rain will percolate into the ground but will appear as springs at the river bed level. Where, in any of the examples given, there is vegetation on the land, either grass or forest, the rainfall is inclined to have a lower run-off percentage (as an immediate run-off into the streams). These conditions, however, may prevent river floods by retaining longer the water on the surface and thus slowing down the rate of its movement into the streams and rivers. It is for this reason largely that deforestation aggravates floods and erosion.

The effect of heavy snowfalls and ice formation from the water vapour evaporated from the oceans appears to be noticeable in the level of the Pacific and Atlantic oceans. Along the Atlantic coast the level is claimed to be lower in the winter than in summer, while the reverse is said to be true in the case of the Pacific. In the former case there is a considerable snowfall in winter followed by an equally considerable flow of water back to the Altantic in the summer. In the case of the Pacific there is considerable evaporation in the summer and a smaller return of water in the winter. It has also been stated that the overall effect in both oceans is that the Atlantic level is rising 0·1 ft. every twenty-five to thirty years, and the level of the Pacific is rising 0·1 ft. in every forty to fifty years. If these statements are based on reliable observations then we are probably concerned with the thaw at the closing stage of the Ice Age which was at its greatest in Pleistocene times in Northern Europe and North America. The melting of enormous quantities of ice would appear to be on explanation, perhaps a part of the explanation, while submarine springs discharging subcrustal water may be an additional factor. In this connection it may be remembered that the Mediterranean is kept filled (against its losses by evaporation) by a current from the Atlantic through the Straits of Gibraltar, and that the Red Sea is similarly replenished by a circulation from the Gulf of Aden. In both these cases the surface and bottom currents are measurable, but in the Labrador current measurement is complicated by the Gulf Stream in the North Atlantic, while Kuro Siwo Drift is a great eddy in itself in the North Pacific. There is almost continuous sea all round the Antarctic about latitude 60° south, so that any measurement would be well nigh impossible.

Engineers are interested in the run-off flow into streams rather than in the academic aspects of the oceanic replenishment, and so elaborate measurements are taken almost daily and annually for determining the volume of water carried by important streams and rivers. Stream gauging is now an accepted routine, and the results even of fairly well known rivers are often of great interest and value. Taking the case of the Columbia River, the largest river flowing into the Pacific from the American continent and one which is one of the greatest hydro-electric power yielding streams, gauging its flow has proved of a necessity. It is 1,200 miles long, begins among the icefields of the Rocky Mountains, and drains nearly 260,000 square miles of territory. It descends 1,000 feet in 400 miles in the State of Washington. Its discharge at Portland (Oregon) into the Pacific Ocean, reckoned as the mean flow, is 280,000 cubic feet per second. The maximum discharge is recorded as about 1,250,000 cubic feet per second. This is about equal to the Nile in full flood at Aswan in August–September when the Nile carries about 280,000 cubic feet of water per second. The Nile discharge above Aswan in 1893 was as follows :—

	Cubic Metres per sec.	Flow Metres per sec.
May	650	0·665
June	550	0·505
July	2,400	0·940
August	7,500	1·720
September	8,000	1·700

The volume of the flood at Aswan between the 1st July and 31st October has varied from 80,000,000,000 cubic metres in 1878 to 41,000,000,000 in 1902. The mean is roughly 65,000,000,000 cubic metres for the same period. (To convert cubic metres to cubic feet multiply by 35·3107, and, vice versa, multiply cubic feet by 0·02832.)

Details of the flow and floods of the Nile and its tributaries have been provided by H. G. Lyons, in his great monograph *The Physiography of the River Nile and its Basin* (published in 1905, Cairo). He shows how the Abyssinian tributaries, the Sobat, Blue Nile, and Atbara, send great floods into the Nile as a result of the run-off from the rainfall on the Abyssinian highlands. Furthermore, it is seen how the flood from the Sobat impounds the flow of the White Nile above Taufikla, and the Blue Nile (Bahr el Azraq) also impounds the White Nile above

Khartoum, and thus improves the storage while these tributaries provide the main flood down the Nile, assisted by the Atbara, from Berber to Wadi Halfa and so to the Aswan Dam. The flood of the Blue Nile between 21st and 25th August, 1903, was as high as 9,500 cubic metres per second, and for the period between July and October up to 62,761,800,000 cubic metres, which is only a little less than the Nile flood (mean value) at Aswan on the main Nile. The floods of the Nile and its tributaries usually lead to prosperity, and these floods are transfusions which give life and wealth to Egypt, and have done so for centuries. River floods elsewhere are a source of danger and can do enormous damage by overflowing their banks or even changing their course. In India the Damodar River, which is under consideration for multiple projects on the lines of the Tennessee Valley Authority, has been a constant danger to the rich plains of Bengal. Although only 368 miles long and draining barely 5,000 square miles where it enters the plains of Burdwan, it may carry a flood approaching 600,000 cubic feet per second. It joins the Hughli River just above the dangerous shoal of the "James and Mary" sands, some miles below Calcutta, but it was once at Tribeni, some miles above Calcutta, and very nearly changed its course in 1941 when it washed away the road and rail communications between Calcutta and Burdwan. Its channel opposite Burdwan is not capable of discharging more than 300,000 cubic feet per second, so that the additional water floods the adjacent country. Floods of a dangerous type follow rainfall precipitations of 4 inches in twenty-four hours on its catchment in Bihar. The river is swollen by precipitations of $1\frac{1}{2}$ inches in twenty-four hours. The maximum known is 6 inches on its upper catchment, and 10 inches spread over part of the catchment (falling in twenty-four hours).

In arid regions, where heavy rain may occur in a few hours, the resulting floods may not only do considerable damage to property but also flow away so rapidly as to fail to soak into the ground to replenish the underground water for the wells on which the area depends throughout the year. Storage of excess run-off water appears to be recognized as the solution to problems of floods, as the artificial lake or reservoir so formed is a regulator to the stream flow. In this connection it may be mentioned that Lake Tana, in Abyssinia, in the upper part of the Blue Nile (Abbai), has a storage capacity of 30 cubic miles or twice the amount that is carried by the Blue Nile to the Makwar Dam, near Sennar, but only 1·5 cubic miles of water overflow from Lake Tana annually. It is thus evident that the Blue Nile receives barely a tenth of its flood waters from Lake Tana and its headwaters. Placing a dam across the outfall of Lake Tana would increase its storage capacity but could not alter appreciably the floods down the Blue Nile. It would require a failure of the rainfall on the highlands of Abyssinia to affect the run-off flow which passes down the Sobat, the Blue Nile, and the Atbara to influence the " Nile floods " into the reservoir at Aswan and so affect Egypt. It might be claimed, with more logic, that the Makwar Dam, near Sennar, is a far greater threat than any impounding of Lake Tana could be.

In discussing the subject of run-off flow from heavy rains or the discharge of flooded rivers into flat country the question of control remains one of the most pressing in all countries. The Hwang Ho or, as this " Yellow river " is called, " China's Sorrow " has caused immense destruction in the alluvial plains of north China. The river surface is 15 feet above the plains, so that—should the stream break out—its potentialities for damage are enormous. It has changed its course and now discharges into the Yellow Sea south of the Shantung peninsula, instead of into the Gulf of Chihli to the north as it used to do. The Mississippi is a source of anxiety also, and has been the cause of immense loss in its overflooding. At Cairo (Illinois), at the confluence with the Ohio, high flood may go 50 feet above that of low water, and in 1927 the flood rose 56 feet and flooded 28,000 square miles of rich country and rendered 600,000 people destitute. The average discharge of the Mississippi into the Gulf of Mexico is reckoned in round figures as 150 cubic miles a year. However, in the case of disastrous floods, it is not the total discharge but a local accumulation of waters. This occurs in many large and small rivers with serious damage to the population. The Columbia River and its tributaries have been responsible for inundations and wash-aways like so many other rivers. The illustration (Photograph XVII) shows the flooded valley of the Severn in England.

The estimation of run-off flow can never be more than approximate unless gaugings of streams have been made for a period long enough to arrive at averages. Such data are available for rivers like the Nile and its tributaries, but as time goes on the waters are utilized for irrigation and do not pass back to the sea. On page 31 estimates have been given of the discharge of twenty great rivers, and it is seen that their annual discharge averages somewhat over 2,500 cubic miles, and an estimate of 5,000 cubic miles was thought a fair computation for the total run-off from all the rivers flowing into the sea. This, of course, excludes ice and snow slipping off as floes and bergs into the sea, as from the Antarctic and Greenland. The total rainfall on the land was computed at about 24,000 cubic miles (23,903) exclusive of snow (6,060 cubic miles), so that if the computation for the river discharge is not too low at even 6,000 cubic miles annually into the sea the run-off flow is barely 25 per cent. This would appear to be a general average in spite of the fact that some engineers assume, as a rough rule, one-third of the rainfall as loss by evaporation, one-third run-off flow, and one-third loss by percolation. As is easily understood, local conditions affect these questions, and what may be true under monsoon conditions in certain parts of India may be misleading in other countries—tropical or temperate or frigid in climate, and either arid or wet in character.

The differences in run-off may be studied from available information on rivers such as the Missouri-Mississipi, the Amazon, the Nile-Kagara, the Congo, La Plata, the Ob or Obi, and the Lena. All these drain catchments of over a million square miles, but their discharges range from 1,000 cubic miles of the Amazon, 250 cubic miles from the Congo, to 150 cubic miles from the Mississippi, and barely 30 cubic miles from the Nile (at Aswan). It is true the Amazon has nearly twice the catchment of the Congo, but its discharge is four times greater because it enjoys a larger rainfall in spite of the fact that both rivers drain equatorial regions. The Amazon flows east to the Atlantic but has its sources in the Cordillera of the Andes in Peru and Ecuador among snow and ice at heights of 20,000 feet The Congo flows west into the Atlantic from the highlands of Central Africa in the Belgian Congo from lakes and slopes barely 6,000 feet above sea-level. The Obi flowing north from latitude 50° N. to the Arctic Circle carries less than half the volume of the Mississippi which flows south from 45° N. to 30°, although both have roughly the same drainage area and are both in the Temperate Zone, one on the northern part and the other in the southern part.

The Yangtsi Kiang, which drains eastward along latitude 30° N., although of the same catchment area as the Mackenzie, which flows northward from 55° N. on to the Arctic Circle, carries more than three times the discharge of the latter river. The Danube, with a catchment area slightly smaller than the St. Lawrence, which drains the Great Lakes, has twice the discharge of the latter river although both flow eastwards at about latitude 45° N. The Murray-Darling, with a catchment of 250,000 square miles and flowing south-west to the estuary of Lake Alexandrina, connecting with Encounter Bay, rarely carries any appreciable discharge, although it is entirely in the Temperate Zone between 26° and 36° south latitude. And so cases of differences may be given for smaller rivers. The run-off flow from the Shari in the western Sudan flows into Lake Chad which has shrunk to a small spread of water 24 feet deep when flooded by the Shari but otherwise from 20 to 3 feet. Lake Ngami is now practically a marsh and often dry although fed by the Okovango and other rivers. In the case of Lake Chad (40° north latitude) evaporation has been the great factor of loss. In the case of the Ngami (20° south latitude) percolation and diminishing rainfall have been factors, but there have been changes in the catchment by warping whereby the drainage has been altered. Lake Ngami once covered 50,000 square miles and was supplied by the Okovangi, the Chobi, and Zambesi. Now even the Okovangi flows uncertainly into the Ngami depression.

Percolation and Infiltration

A percentage of the rainfall sinks into the ground by absorption, infiltration, or percolation. The word "absorption" is most commonly used of the rain absorbed by plants or vegetation in general, and sometimes by the soil. The common word is "percolation" for the water that sinks down into the rocks through the soil or other covering. The word "infiltration" is usually applied to those cases where percolating water emerges underground, for example, in a well. Whatever nomenclature may be adopted it is dependent on the surface on which rain falls as to what percentage is likely to enter the ground. The factors which affect the disposal of rain as percolation are : the nature of the surface, whether it is forest covered, grassy, cultivated, bare of soil, or with exposed impervious strata ; whether the surface is hot and dry, or cool and moist or wet ; the degree of slope, the humidity of the air, and the rate of precipitation of the rain. In the case of snowfalls the thaw will probably allow the water a better opportunity for percolation as the snow (or ice) remains longer on the surface. It may happen that a heavy and prolonged rainfall, say 5 to 8 inches in twenty-four hours, or 8 to 12 inches in forty-eight hours, falling on almost any surface, even on loose, coarse sand, may so soak the top stratum that it becomes impervious to further percolation, and the remainder of the rainfall then merely flows away as run-off flood water.

The importance of the ground surface, the air temperature, and the rate of precipitation of rain is appreciated in all questions of percolation of water into the ground, but the porosity of the rocks, the disposition of joints and fissures, and the structure of the strata greatly influence the movement of underground water. Loose sand under deep water may be compressed by the weight of the water and the percolation or leakage may thus be less than was anticipated. A heavy earth dam founded on sand under conditions of emergency or necessity may so press on the sand that the leakage under the dam may be relatively small. Much, of course, depends on the size of the grains of sand, since coarse sands allow easier and more rapid percolation than fine sand. In his book *Hydrology and Ground Water*, J. M. Lacey has given the following rates of infiltration to a bore-hole 6 inches in diameter in sands and gravel :—

Fine sand	about	4,000 gallons a day
Medium-grained sand		30,000 ,, ,,
Coarse sand		80,000 ,, ,,
Gravel free of sand		500,000 ,, ,,

Such figures are interesting for general comparison but, for reference, information is required on the size of sand grains, the thickness of the water-bearing strata, and other particulars as to "head" or gravity flow. Charles S. Slichter has provided useful information of this kind

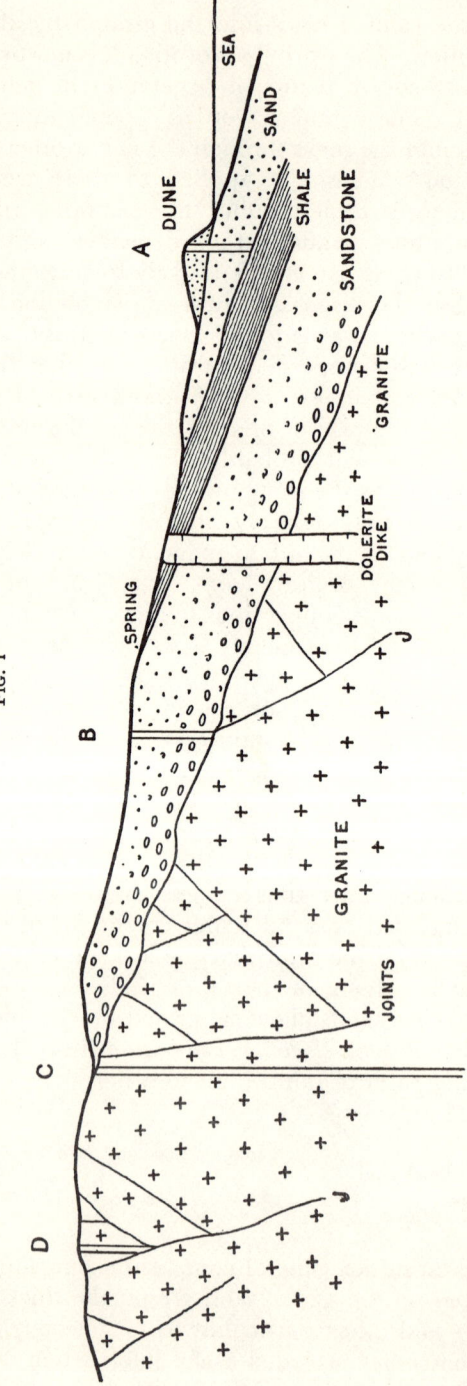

Fig. 1

GEOLOGICAL SECTION SHOWING COASTAL PROBLEMS OF WATER SUPPLY

A. Fresh water found under sand dune.
B. Water held up by Dolerite dike.
C. A failure in Granite.
D. Success by striking point planes.

in his paper on " The Motions of Underground Water " (see *Bull. No. 67*, 1902, Water Supply and Irrigation Papers, U.S. Geol. Surv.). On p. 30, in Table iv, he gives the following details :—

Kind of Soil.	Effective Size of Grains. mm.	Velocity in in. per mile, Grad. 1 : 1.	Velocity in miles per year, Grad. 1 : 1.	Velocity in miles per year, Grad. 100 ft. : mile.
	0·01	0·0014	0·0113	0·00026
Silt	0·02	0·0054	0·0452	0·00102
V.F.	0·05	0·034	0·2823	0·00638
Sand	0·10	0·1361	1·129	0·02551
Fine	0·15	0·3063	2·541	0·05753
Sand	0·20	0·5446	4·518	0·1021
Medium	0·40	2·178	18·07	0·4081
Sand	0·60	4·901	40·65	0·9183
Coarse	0·80	8·714	72·28	1·633
Sand	1·00	13·61	112·9	2·551
Fine	2·00	54·46	451·8	10·21
Gravel	4·00	217·8	1,807·0	40·81

In this connection it is to be noted that pore space volume is frequently greater the less porous or more impervious a material is to the passage of water through it. The following values of pore space volume are supplied for examples of soils : 32·5 per cent in sandy loam, 34·5 per cent in loam (a mixture of sand and clay), 44·1 per cent in heavy loam, 47·1 per cent in clay loam, and 52·9 per cent in heavy clay. Here, the clay with greatest pore space volume (more than half the volume of the material) is the most impervious substance for percolating water. W. R. Baldwin Wiseman, in his paper " The Flow of Underground Water " (see *Proc. Inst. C.E., Minutes*, vol. clxv, 1906, Paper No. 3594) gives the rate of flow through filter beds :—

Filter Bed at—	Discharge per sq. ft. per hour. Gals.	Cu. ft.	Thickness of Filter Bed in Feet. Fine Sand.	Coarse Sand.	Height of Water on Bed, in ft.	Total Head, in ft.
Birmingham	2·34	0·375	3·00	3·6	2·00	7·50
Bradford	2·30	0·369	0·50	4·50	3·50	11·00
Cheltenham	2·45	0·393	3·50	4·50	1·50	6·00
Derby	2·85	0·456	2·50	3·50	2·00	7·50
Exeter	1·21	0·194	0·50	3·50	2·50	6·75
York	1·07	0·171	4·00	2·50	2·50	9·00

Details are not provided of the effective size of the grains of fine and coarse sand, but it seems clear that in the beds at Exeter the material must be on the fine side. This also applies to the filter bed at York.

H. J. Llewellyn Beadnell has given an interesting record of the interference of one well, A, on another well, B, at a distance of 1,860 feet or so. The well A stopped pumping at 7 p.m. on the 12th June, 1907, and after a night's rest was started again at 7 a.m. next morning, 13th June, pumping at 114 gallons per minute. The other well continued pumping steadily but its discharge was definitely affected, as seen :—

Time.	Discharge from B. gals. per min.	Time.	Discharge from B. gals. per min.
7.00 p.m.	61·2	7.00 a.m.	83·7
8.15 ,,	65·6	8.00 ,,	78·4
9.00 ,,	68·4	9.00 ,,	75·0
10.00 ,,	69·6	10.00 ,,	73·0
11.00 ,,	73·2	11.00 ,,	70·80
12.00 ,,	74·7	12.00 ,,	69·60
1.00 a.m.	76·6	1.00 p.m.	69·0
2.03 ,,	77·4	2.00 ,,	67·70
3.00 ,,	79·2	3.00 ,,	66·8
4.00 ,,	79·7	4.00 ,,	66·6
5.00 ,,	82·1	5.00 ,,	66·2
6.00 ,,	83·1	6.00 ,,	65·3
7.00 ,,	83·7	7.00 ,,	64·00

It is necessary to explain that well A is at a level of 5·9 feet below the outlet from well B; that well A is 646 feet deep and penetrates 311 feet into a water-bearing sandstone; well B is 479 feet deep and penetrates 200 feet into the same water-bearing sandstone (which is evidently inclined or dips towards the deeper well A. The diameter of the pipe at A is $5\frac{5}{8}$ inches and that of B is 8 inches. There was no evidence of any other cause of interference except the effect of pumping A and the resulting influence on the capacity of B.

W. Whitaker has recorded in his report on " The Water Supply of Surrey " (*Mem. Geol. Surv. England and Wales*, 1912, p. 78) that the effect of interference by wells has been a frequent subject of controversy but he has quoted two examples as proof. One in the case of a London well into the sandy stratum above the chalk, where a well in Thames Street was affected by a well in Southwark in 1842 (see *Proc. Inst. C.E.*, vol. ix, 1850, p 169). These wells would be on opposite sides of the river. The other refers to a well near Sutton (by the London and Brighton Railway Company) which affected important springs at Carshalton, and quotes W. V. Graham's paper on the subject (see *Trans. Inst. Surveyors*, vol. xxxix, 1907, part ix, pp. 330–1). He also gives an example of how a tunnel caused a diversion of the flow of underground water from one drainage area into another. The case was that of the Sodury tunnel (Great Western Railway) through the Jurassic escarpment of the Cotswolds (and also in the scarp of the Lower Greensand where the Sevenoaks tunnel went through in Kent). The underground water was drawn from the dip side (inward from the scarp face) and brought out by the tunnel in the opposite direction (to the scarp side). And yet again, in the case of the Merstham tunnel, when so much water was encountered that a lower drivage or adit was made to draw off the water. The springs in the neighbourhood were drained and the Merstham Mill Head was dry for some years until the water from the adit was turned into the Old Mill Head.

Run-off rain may pass underground in large volumes in country where the rocks are fissured or contain solution cavities. Limestone regions are well known for their caves and " swallow holes " and it is not an uncommon phenomenon to see large streams disappear into the ground in such areas. Among " show " places of the *dolinas* and *grotto* of the *Karst* or *Carso* (limestone) country along the north of the Adriatic perhaps the most interesting are the grotto of St. Canziano, east of Trieste, into which the River Timavo disappears, and the cave of Adelsberg, near Postumia (in the same region), into which the Puika river flows underground. Neither river reappears, but it is believed that both discharge as subterranean springs into the Gulf of Panzano, north-west of Trieste. However, the examples quoted provide evidence that quite an appreciable amount of the rainfall which percolates into the ground re-emerges and flows into the rivers or directly into the sea. Subterranean springs are often of great size, such as those around the coast of Greece (in the gulfs of Argos and Corinth, etc.), and of Italy (the Gulf of Spezia and that of Taranto). Springs of a similar freshwater type are found in the Persian Gulf at the Bahrein Islands, in the Gulf of Muscat, near Muscat, and along the south coast of Arabia about Dhufar in Oman. There are artesian springs along the West Australian coast, near Perth, and even on the borders of the dried-up Lake Eyre, but these are quite distinct and of deep-seated origin.

The estimation of percolating water is not possible without allowances for many factors, which it is practically impossible to determine with accuracy. Locally it is a simple matter to ascertain the rainfall and to gauge the streams, and so ascertain what proportion of the rainfall is run-off and then to make an allowance for evaporation and so determine the percolation by difference. In such a procedure it is understood that several considerations are to be noted. One is that a portion of the percolation is absorbed by the soil and by vegetation, and is subsequently transpired from the plants or evaporated from the soil and so returns to the air. Another detail is that some of the water which infiltrates reacts with the minerals in the rocks and hydrates them, and so becomes held in combination. A third factor is that some of the water percolates down along fissures and becomes heated and dissolves matter from the rocks, and again ascends. This ascending water will deposit its dissolved matter as it cools, either in rock fissures or in the interstices of the strata, and so forms mineral deposits, on the one hand, or cements the porous beds, on the other. The ascending water may emerge as mineral springs or be held below ground under artesian conditions. The circulation of the water of the earth's surface, from the oceans back to the oceans, is in the main an annual cycle, but the water given up by thermal springs may be of longer duration although it is part of the original percolation which was obtained by the difference between run-off and evaporation from rainfall.

Weathering and Hydration

There is considerable difference of opinion among geologists as to the depth to which infiltrating water may percolate. Some consider that water will penetrate down to depths where the rocks are still strong enough to allow fissures to persist, and reckon 32,000 feet or roughly 6 miles as this limit. After that the heat and pressure are great enough to close fissures and all but capillary pore spaces and so prevent the descent of water, except in a state of combination with the mineral or rock substance. Long before the depth of 32,000 feet the percolating water will have given up the constituents, oxygen, carbonic acid, nitric acid, etc., which it has brought down from the atmosphere as rain. These dissolved components in rainwater make it a powerful solvent and in this form, as meteoric water, the percolating liquid attacks the rock minerals and dissolves the soluble matter which results from these reactions. The processes are known as weathering, since they take place from the surface to shallow depths, or within the zone that rainwater still functions as an agent of direct attack. Below the depth or zone of weathering, which may vary from less than 25 feet to depths exceeding 250 feet, depending on the configuration of the surface, the percolating water will have changed and will carry mineral matter in solution. As this percolating water continues downward it enters the zone where it may react again and alter the rocks with which it comes in contact, or it may deposit mineral matter in the pore spaces of the stratum through which it travels, and cement them. This lower zone, extending to great depths, is referred to as the zone of cementation.

If fissures are available the infiltrating water may continue in the meteoric form to great depths, and may follow channels rather than percolate. The reactions from such waters will be the same as those in the zone of weathering. That is, oxidation and solution and hydration of the minerals may occur, and possibly some deposition, by replacement of one dissolved substance for another more soluble in the altered water. The reactions in the zone of cementation still involve water in its liquid condition, and it still functions as a solvent from which deposition may occur or through which replacement phenomena take place. Some idea of the temperatures and pressures involved at great depths in the earth's crust are shown in the data below :—

Zone: Surface Rock, 2·6 sp. gr.	Depth in Feet.	Pressure in lb. per sq. in.	Temperature in ° C.
Sub-soil	1·0	1·12	0
Weathering	328·0	367·3	3·3
Cementation	3,280	3,673·6	36·7
High pressure	32,806	36,736·0	367·3
High temperature	65,612	73,472·0	734·7
Magma	131,224	146,944·0	1469·5

At a depth of 3,280 feet (1,000 metres) the pressure is above the critical pressure of steam, while at a depth of 32,806 feet (10,000 metres) the temperature is almost that of the critical temperature of steam. Below a depth of 32,806 feet the subject of underground water passes into the sphere of speculation. Weathering (absorption of water into some combination with the rock minerals) occurs at shallow depths and relatively ordinary temperatures. Hydration as part of the deeper zone of weathering occurs within depths of 328 feet (100 metres) and at temperatures which may be warmer than temperate climates and cooler than tropical temperatures provided no strong exothermic reactions occur. Water is actually absorbed by hydration and becomes unavailable, whereas in the deeper zone of cementation water remains uncombined. There is thus loss in the percolation of water through the zone of weathering to that of the zone of cementation.

Hot Springs and Volcanoes

There appears to be little doubt that many hot or thermal springs are due to the return of meteoric waters to the surface, after they have travelled down, either by percolation or as interstitial water which had been trapped during the deposit of silts and sands and related sedimentary strata. In both cases the percolating or trapped water has been carried down to depths greater than 10,000 feet where temperatures of 100° C. are normally expected (and where the weight of the strata may be 1,120 lb. per square inch (half a ton per square inch). In other cases, as in geysers and natural discharges of steam, and other heated gases, there is some uncertainty as to the history of the water. It may be of meteoric origin or may be water given up from the rock magma for the first time, " virginal water." The same remarks apply to the steam that is discharged in such vast quantities from volcanoes. In many instances, seeing how closely volcanoes follow orographic axes (mountain ranges) through islands or along a coast, there is a strong suspicion that sea water has played a part in the fusion and subsequent eruptions of lava. It may even be possible that " trapped water ", as in the hygroscopic (held by colloidal matter) and capillary (held in the pore spaces of clays) water, in argillaceous sediments, has been carried deep into the earth's crust and played a part in the fusion of the rock containing it. With the fusion and subsequent escape of the molten lava the contained steam has also escaped, rejoined the water vapour in the atmosphere, and then as rain fallen back into the sea. The circulation of such "trapped" waters, whether by hydration, capillary, or other mode of combination in the rocks, may have a cycle of millions of years from the time the water fell as rain from clouds and entered the ground as percolating water to the time the water fell again as rain after a volcanic eruption and returned to the ocean.

THE NATURAL HISTORY OF WATER

Summary of Rainfall disposal

In the previous paragraphs of this chapter the cycle of water on the earth has been estimated, and it has been found that while there are, so to speak, "fixed" deposits of moisture or water in the atmosphere, lakes, and rivers and in the rocks, yet by far the greatest reservoirs for water are the oceans and seas. From these sources water vapour is carried into the air and precipitated as rain, which is partly re-evaporated, partly runs off into the streams and rivers, and partly sunk into the ground to percolate downwards, or infiltrate into the rocks for their hydration. These estimates show :—

(a) The evaporation from the oceans averages annually about 53,000 cubic miles of water ;
(b) The evaporation from the lakes and rivers is computed at roughly 378 cubic miles of water ;
(c) The evaporation from snowfields and glaciers averages, annually, 6,048 cubic miles of water.

The total in round figures is taken as 60,000 cubic miles. It is presumed that all this water vapour is re-precipitated as rain and snow, and the estimates of the disposal of this precipitation have been computed as follows :—

(d) The rainfall on 42,000,000 square miles of land surfaces yearly is 23,903 cubic miles ;
(e) The snowfall on 15,000,000 square miles of land surfaces is about 5,681 cubic miles ;
(f) The rain and snowfall on 140,000,000 square miles of ocean by difference, is 59,426 cubic miles.

The total in round figures would be 60,000 cubic miles in the proportions—24 : 6 : 30 millions of cubic miles.

As regards the disposal of the above 60,000 cubic miles of annual rainfall, of which a half, 30,000 cubic miles, falls as snow and rain on the land, it is estimated that :—

(g) The run-off flow into the rivers, as measured by their discharge into the oceans and seas, amounts to roughly 6,000 cubic miles, 20 per cent of total land snow and rainfall and 25 per cent of the rainfall alone.
(h) The evaporation from the land (from lakes, rivers, and snow and ice) has been estimated under (b) and (c) as roughly 6,500 cubic miles, but this does not include an enormous re-evaporation of rainfall (including transpiration from plants, and other water vapour). It is practically impossible to estimate the evaporation from rainfall on an average basis, but so high a figure as 50 per cent of the rainfall may be so lost.

Photo, Chambers. The National Geographic Society.

PLATE XI.—THE GRAND CANYON OF THE COLORADO RIVER IN ARIZONA.
The view shows perhaps the greatest visible example of river cutting or stream erosion in the world. The photograph is a view of the inner gorge looking downstream from the south rim. This part of the valley is the Grand Canyon National Park and so is preserved as a place of natural remarkability.

[*Facing page 64*

PLATE XII.—THE NIAGARA FALLS.
The view shows the " American Falls " on the left and the " Horse Shoe " (Canadian) Falls beyond.

Public Relations Dept. S. Rhodesia.

PLATE XIII.—VICTORIA FALLS, ZAMBESI RIVER, RHODESIA.
Visited by Their Majesties the King and Queen in April, 1947.

[*Facing page* 65

(*i*) The subject of the disposal of the rainfall as " loss by percolation " is similar to that of evaporation under (*h*) (above). The average by difference would appear to be 25 per cent of the rainfall, but this proportion may be greater and that of evaporation less.

Among civil engineers it is a common rule that the rainfall is disposed of in equal proportions as run-off flow, loss by evaporation, and loss by percolation. Locally, these proportions can be determined by stream gauging, rainfall records, and determination of evaporation by keeping records over a period of years.

In his book *The Realm of Nature* (1932), Hugh Robert Mill has stated (p. 267) that observations of the River Thames, for a catchment of 3,850 square miles and rainfall of 28 inches above Teddington Weir, showed that the run-off was 0·55 cubic miles out of a total of 1·70 cubic miles. This is roughly a third. The remaining 1·15 cubic miles was credited to evaporation and percolation and some movement through the gravel under the river bed. However, these details are fundamentally affected by various factors—climate, rate of rainfall precipitation, nature of surface (slope and soil), etc. Where the rainfall is low, say 10 inches a year, and the catchment is *good* (steep and impervious), the run-off may be a million cubic feet per square mile. If the catchment is *bad* (a flat surface of porous soil), the yield may not be half a million cubic feet per year. A rainfall of 20 inches a year on a *good* catchment should yield a minimum of seven million cubic feet (run-off flow), wheras on a *bad* catchment the anticipated yield might be less than four million cubic feet optimistically. (See the author's *Geology of Water-Supply*, 1949, pages 42–3.)

PART 2

THE WORK DONE BY WATER

PART 2.—THE WORK DONE BY WATER

Chapter IV.—Erosion of the Land Surface

In the previous chapters an effort has been made to show the properties of water in its normal liquid state and also in the gaseous form of vapour and the solid form of snow and ice, and to outline its circulation under the influence of the sun's heat by evaporation from the seas, precipitation as rain, and percolation into the ground. Falling rain scours the surface. Frost causes rocks to split by the expansion that water undergoes in its conversion to ice. Landslips are caused by lubrication as a result of rain or of percolating water. Transportation of detritus is effected by running water. Erosion is often the result of coastal currents and waves. Finally, man is himself an agent in reducing or increasing erosion or denudation by deforestation on the one hand or revetting to prevent slips on the other, and in several other directions as a result of engineering projects.

Denudation by Rain and Rivers

The direct effects of rain depend on the surface on which it falls, the humidity of the region generally, and the rate of precipitation of the rain. On flat surfaces of alluvial soil, which may be soft and porous or stiff and impervious, the rainfall may sink into the ground and saturate the marl or sandy alluvium, or it may wet the clayey deposit and then run off its surface as well as go back into the air by re-evaporation. If the alluvium is above drainage level the water soaking into it will emerge and join the streams, and thus drain the porous material. In the other case, the absorption of water by the clay may result in causing it to soften and slip down and so get washed away. Where the rain falls on hard rock it may penetrate fissures and widen them. In times of frost the moisture within such cracks will freeze and result in angular pieces of the rock breaking off and adding to the scree below. Where forest or other vegetation covers the surface the rainfall does not strike the soil or rock directly but may soak into the heavy soil on steep slopes and help any " creep " to develop into landslips. In the case of very heavy falls of rain, all soil may be washed away and the bare rock exposed to scouring. In the case of limestones and some other relatively soluble rocks, the wash of heavy rain will carve into the rock by taking away softer and more soluble matter and will leave a cliff face carved and even form hollows and caves. An example of this kind is to be seen in the cliff facing the Damascus Gate at Jerusalem, worn to resemble the face of a human skull.

Where the rain flows over the surface, first in rivulets and then into ever larger streams, the running water scours the ground and, with dislodged particles, cuts into the ground, deepening its channel and removing more and more material. Perhaps the most spectacular example of stream erosion is that of the Grand Canyon of the Colorado River in Arizona. The following extracts give a good description :
" It ranges in width from 4 to 18 miles, its greatest depths lie more than a mile below its rim and it extends in a winding course from the head of Marble Gorge, near the northern boundary of Arizona, to Grand Wash Cliffs, near the Nevada line, a distance of 280 miles. Its most impressively beautiful part, 56 miles long, lies within the Grand Canyon National Park. . . . The Grand Canyon was formed by the ceaseless cutting of the silt-laden Colorado River . . . it has cut its way vertically downward, maintaining its course almost without change. Meanwhile, the rocky walls of the canyon have been exposed to destruction by the action of rain and rill, of frost and landslide, of wind and chemical action." It represents an excavation of some 740 cubic miles of rocks. The rocks exposed in a north and south section, from Tiyo Point (7,765 feet), on the edge of the Kaibab plateau, to Hopi Point (7,071 feet), at the rim of the Coconino plateau, and down to the Granite Gorge (2,350 feet) between them, are gently dipping and relatively soft Palæozoic limestones and shales and sandstones, overlying, with a slight unconformity, gently inclined pre-Cambrian strata which, in turn, lie quite unconformably, upon Archæan gneisses and schists (which are seen in the river exposures). It is thought that this amount of excavation has been done within geologically recent times (Pleistocene), say, within 1,000,000 years. It means the removal of 750 cubic miles of rock in that time, or the transport of roughly 7,500,000 tons a year.

The transporting power of a water current is reckoned to vary as the sixth power of its velocity. According to the law governing the transporting power of a fluid, the diameter of the material, or object of similar shape and density, which is carried increases proportionally to the square of the velocity of the fluid. Thus, while a current of water, travelling at 6 to 9 inches per second, will transport fine sand (grains 0·01 inch diameter), it will require a fresh breeze of 12 miles an hour (18 feet per second) to blow away the same material. In the case of coarse sand (grains 0·04 inch diameter), a water current of a foot a second will carry the material, whereas it can be wind borne only by a strong wind of 25 miles an hour (36 feet a second). For larger material, say, gravel of 1 inch diameter, it would require a current of water at $2\frac{1}{2}$ feet per second (say, $1\frac{3}{4}$ miles an hour) to carry it, while a hurricane of 100 miles an hour (a very rare velocity for wind on the land) could hardly move the gravel. R. D. Oldham has given a brief description of the carrying power of a flood stream in the

Cherrapunji (Assam) region which is subject to heavy rain. He wrote: "... The water had risen only 13 feet above the level at which it had stood a few days previously; the rush was tremendous—huge blocks of rock measuring some feet across were rolled along with an awful crashing, almost as easily as pebbles in an ordinary stream. In one night a block of granite, which I calculated to weigh upwards of 350 tons, was moved for more than 100 yards; while the current was actually turbid with pebbles of some inches size, suspended almost like mud in the rushing stream. ...." In that region there now is practically no soil on the Cherrapunji plateau, and it is also noticeable that water carrying much mud in suspension (and its increased density therefrom) carries larger stones than clear water, for equal velocities.

Since the size of materials is seldom clearly stated, it is useful to have a quantitative scale as follows:—

Mesh of Sieve Apertures. per in.	Apertures in— in.	mm.	Class of Material.
5	0·10	2·54	Fine gravel
8	0·062	1·574	Coarse sand
10	0·05	1·270	,, ,,
12	0·0416	1·056	,, ,,
20	0·025	0·635	Medium sand
30	0·0166	0·421	,, ,,
40	0·0125	0·317	,, ,,
50	0·010	0·254	Fine sand
70	0·007	0·18	,, ,,
100	0·005	0·127	,, ,,

The number of the mesh is according to the grading of the Institution of Mining and Metallurgy and written IMM. Other data of this kind are shown in the following tables:—

WATER CURRENT VELOCITIES FOR SCOUR (after Dubuat, 1865)

Material.	Velocity in in. per sec.
Fine potter's clay	3
Fine sand	6
Coarse sand (size of linseed)	8
Fine gravel	12
Pebbles (1 in. diameter)	24
Angular stones (size of an egg)	36

And to define further the terms used for similar materials the following scouring velocities are added:—

Material.	Velocities in ft. per sec.	m.p.h.
Soft clay	0·25	0·17
Find sand	0·50	0·34
Coarse sand and gravel	0·70	0·48
Gravel (size of French bean)	1·00	0·68
Gravel (1 in. diameter)	2·25	1·54
Pebbles (1½ in. diameter)	3·33	2·27
Heavy shingle	4·00	2·73
Soft rock, brickwork, etc.	4·50	3·08

W. M. Griffith has discussed the theory of silt and scour in a valuable contribution (*Proc. Inst. C.E.*, vol. 223 (1927)); he considers that the silt transporting power of a stream varies with the differences in the velocities in different parts of the current, and states that the laws governing silt transportation and silt deposition are the same, and that the law of scour is not governed by the law of silt transportation on slopes steeper than the natural slope of saturated alluvium because the particles tend to roll down the slope under gravity; and, according to him, Parker's table (see below) gives bottom velocities above and below which the material indicated is carried or settles in water:—

Material.	Velocity in ft. per sec.	Material.	Velocity in ft. per sec.
Soft earth	0·25	Gravel (½ in.)	1·5
Fine clay	0·25	Pebbles (1 in.)	2·0
Soft clay	0·50	Pebbles (egg size)	3·5
Fine sand	0·70	Stones (3 in.)	4·5
Coarse sand	0·80	Boulders (8 to 6 in.)	6·5
Sand and gravel	1·00	Boulders (12 to 18 in.)	10·0

And J. N. Dignes La Touche has drawn attention to the fact the bed is altered between flood and low water because the rapid flowing flood waters scoop out the bed, which silts up again as the current slackens. He mentions, for alluvial beds, that in high flood the water holds up the bank rather than scours it, but water is absorbed by the dry soil in large quantities. When the water level falls again water seeps back from the subsoil, carrying silt with it. This exuded silt will be carried away as fast as it slips out and so a bank may become undercut and collapse. The result may be a stable slope or it may lead to the river breaking out (see *The Young Engineer*, 1936, p. 102).

In his book, *The Nile Tributaries of Abyssinia* (1867), Sir Samuel W. Baker describes the Atbara River in flood. At first, in June, he found "... the river was dead; not only partially dry, but so glaring was the sandy bed that the reflection of the sun was almost unbearable...". Then he came again in mid-July and says: "... we had suddenly arrived upon the edge of a deep valley, between five and six miles wide, at the bottom of which, about 200 feet below the general level of the country, flowed the River Atbara.... Here was the giant labourer that shovelled the rich loam upon the delta of Egypt. Upon these vast flats of fertile soil there can be no drainage except through soakage. The deep valley is therefore the receptacle not only for the water that oozes from its sides, but subterranean channels bursting as land springs from all parts of the walls of the valley, wash down the more soluble portions of earth, and continually waste away the soil. Landslips occur daily during the rainy season; streams of rich mud pour down the valley slopes, and as the river flows beneath in a swollen torrent the friable banks topple down into the stream and dissolve. The Atbara becomes the thickness of pea-soup as its muddy waters steadily perform the duty they have fulfilled from age to age...." It is estimated that the Nile carries 45,000,000 tons of silt annually into Egypt. An analysis of Nile mud taken in 1895 and analysed in 1896 by A. von John shows :—

	Per cent.		Per cent.
Silica	45·10	Potash	1·95
Alumina	15·95	Soda	0·85
Ferric oxide	13·25	Water (comb.)	6.70
Magnesia	2·64	Water (moist)	8·84
Lime	4·85	SO_3 (sulphate)	0·34

Analyses of the Nile waters are given on page 80.

Further examples of the direct effects of erosion are seen in the earth pillars, each capped with a pebble or fragment of hard stone, which are to be found where the rainfall is very heavy and the underlying strata are soft. Among the more astonishing of such " pillars " are those on the western slopes of the Rocky Mountains, in the Sawatch region of Colorado. Here, some of these " earth pillars " due to the wash of rain, now stand nearly 400 feet high, and represent this depth of removal of solid material. In the case of the isolated, small plateau on the borders of British Guiana and Venezuela, standing at an elevation of about 9,000 feet above sea-level, Roraima and others, it is clear that they are all that is left of an extensive plateau. The capping bed, or umbrella, is a hard conglomerate which overlies soft sandstones. It is the lower strata which have been washed away by the heavy rain except where a mass of the hard bed has protected it. The result is a plateau which stands 1,500 feet above the forest-covered slopes below, and is, indeed, a kind of gigantic earth pillar. As there are several along that water-shed, the action must be considered the same as with a " relict " mountain. Mountains such as the Aravallis of Rajputana, India, still standing to heights of 4,000 feet, are the mere " roots of mountains ". It is impossible to say to what heights the original mountains raised their peaks. All has been worn down to almost base level. Similarly, the Archæan rocks of the Canadian " Shield " are believed to represent regions which have been exposed to denudation for millions of years, and from which miles (in thickness) of rock has been removed.

Waterfalls and Gorges

A description of the Grand Canyon of the Colorado River, in Arizona, has already been given as a spectacular case of extensive erosion by the cutting action of a relatively small river. The case most commonly given for the cutting action of a river is that of the gorge below the Niagara Falls. The falls, American and Horse-Shoe, are over low, upstream, dipping limestone which are underlaid by shales. Scouring at the base of the falls is backwards in the softer stratum, and so the limestone is undermined and the falls retreat. The waters

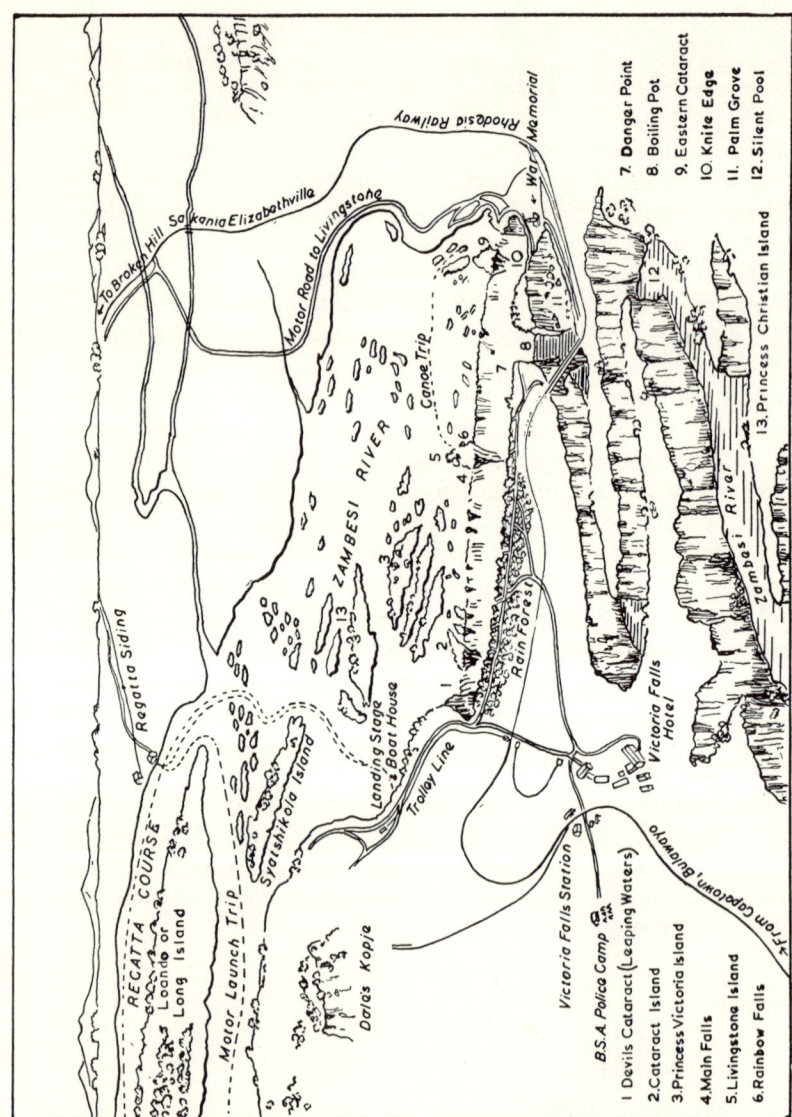

FIG. 2.—VICTORIA FALLS AND GORGE (after H. B. Maufe)

A. Shallow well into sandy alluvium near stream bed.
B. Deeper well after boring proves bed sands.
C. Deep boring into gravel filling an old bed.
D. Shaft to meet stratum which taps the old bed and up which a drivage will go.

are free of silt, or their erosive power would be greater, and Lake Erie would probably have been drained long ago. As matters stand, this action of cutting back or recession of the falls has cut a gorge at least 7 miles long from where the limestone scarp is seen above the gorge (downstream). Estimates suggest that the river is cutting actively, and that the recession of the American Falls is taking place at about 6 inches a year and the regression of the Horse-Shoe Falls is nearly 2 feet a year. Although hydro-electric power is generated from both falls, the diversion of water has been limited, so that the falls at Niagara may be preserved as a " show place ". This diversion of some of the water may also prolong the life of the falls by reducing the erosion that causes their recession. The American Fall is 165 feet drop and 1,000 feet broad, while the Horse-Shoe or Canadian Fall drops 160 feet and is 2,000 feet along its curved width. Computations as to the age of the gorge and the initial falls (7 miles downstream) suggest a period of 25,000 years, which is nearly 18 inches a year. Perhaps of less grandeur and beauty, but nevertheless very remarkable, is the Kaieteur Fall, on the Potaro River, in British Guiana. The river drops over from a plateau 822 feet high in one sheer fall of 781 feet. The width of the water at the top of the fall is 120 feet in the dry season but swells to 400 feet when the river is swollen by rain.

Mention should be made of the Victoria Falls, on the Zambezi River, in Rhodesia. The river water is clear and placid as it flows to the islands of the rim, and then simply disappears in a canyon over 6,000 feet long, barely 200 feet across and 400 feet deep. The water drops the full 400 feet into the chasm followed by a mighty roar and accompanied by a spray which rises up as mist yielding rainbow effects. These falls are therefore nearly twice the width of Niagara and about twice as high, but their grandeur and might are largely hidden by the manner of their drop into the chasm. Along the bed or floor of the chasm, which has been cut in horizontal beds of basaltic lava (of Triassic, Mesozoic age), the churned and swirling water continues down a zig-zag gorge for several miles. Above the falls the waters are clear blue and largely free of sediment, and discharge from 60,000,000 to 100,000,000 gallons a minute into the chasm. The story of " mosi-oa-tunya " (the smoke that sounds) led Livingstone to visit the place and discover these remarkable falls. According to H. B. Maufe the zig-zag gorge below and the chasm at the falls are due to the water scouring along well-developed fissures (joint planes) in the basaltic lavas. These planes of weakness cross the river and so offer scope for special erosion. By this geological study it is possible to foretell that a new chasm will, in the distant future, be excavated, either along the joint (crack) from Cataract Island or near the lip of the Eastern Cataract, dependent on the relative rates of erosion at each place. The geological age of these falls, in the basaltic lavas, is probably since Pleistocene times but older than Niagara, and the recession is probably slower.

The Action of Frost and Ice

Mention has already been made of the expansive force which is exerted when water is frozen, and how the action of frost on the water in rock crevices causes the rock to splinter and so form the scree which is seen at the base of bare rock cliffs. The phenomenon is so well known that there have been cases where the climatic conditions of a past geological age have been suggested by the discovery of breccia-like beds. In some cases these deductions appear to have been erroneous, since it is known that similar scree may form as a result of rocks being heated by the sun all day and then being chilled at night. The expansion and contraction of the rock itself has caused the splitting without any moisture, since the phenomenon is most common in hot arid regions. W. W. Watts has stated, *Geology for Beginners* (1942), p. 261, that some of the Upper Palæozoic, Permian, breccias of England, which contain angular fragments of limestone, " are often screes weathered by frost from limestone cliffs, and consolidated by cement when buried under the succeeding sediments." He also adds, page 32, that " In hot countries . . . the mere alternation of heat and cold has a somewhat similar effect . . . " and that " The alternate wetting and drying of a rock by rain and sunshine has a similar but rather less effect ". It should be stated that the Permian beds of England, with their salt and gypsum and red strata, definitely indicate arid conditions in a hot country such as Russian Turkistan, in the basin of the Sea of Aral of southern Kazakstan. Obvious examples of scree resulting from frost action may be studied in the Lake District of England, say, along the eastern side of Wastwater.

A covering of snow tends to protect the rock below, until the thaw, when the melting snow penetrates into the ground or adds to the flow of streams. Where snowfalls are heavy or accumulate to become ice for glaciers or icefields the weight of the resulting ice and its movement introduces erosion by abrasion. Great masses of ice develop a kind of fluidity which permits the mass to flow down valleys exactly as a river of ice, or to flow outwards over irregularities of continental lands. Part of this movement is due to the phenomenon of regelation, whereby the ice melts under great pressure but resolidifies on release of the pressure. Much has been written on glaciers and ice-sheets. Some famous glaciers have great thicknesses of ice, as much as 1,000 feet, and move slowly at an inch an hour (2 feet a day), as in the case of the Mer de Glace, in Switzerland. The longest Swiss glacier is the Aletsch, about 15 miles long. The Biafo glacier, in Baltistan, Kashmir Himalaya, is nearly 36 miles long. In Scandinavia the snow line is about 4,000 feet above sea-level, in Switzerland (Alps) about 8,000 feet, and in East Africa, at Kilimanjaro, it is 16,000 feet. Where there is sufficient gathering ground above the snow line of any area, snowfields and glaciers occur all over the world. Some of the great glaciers of Greenland travel between 18 to 30 inches an hour (60 feet a day) as the sheets push down into Disko bay to float away as icebergs. The longest glacier known, the Beardmore, in the Antarctic, is 200 miles long and descends 8,000 feet to the Ross Sea (South Pacific). In most cases the water emerging from under a glacier is milky with the mud of abrasion, and the moraines show how fragments are carried by glaciers. It is not so well known that icebergs contain embedded rock, often facetted and striated as a result of grinding along under the moving ice on the land below. It would be very difficult to estimate the amount of detritus carried by glaciers and ice-sheets annually.

The Pleistocene glaciation or Great Ice Age of North America and Northern Europe has left extensive deposits of boulder clay, together with striated rock surfaces, to show how widespread these ice sheets were a million years ago. It is not too much to say that the plains which stretch from near Berlin to Moscow are covered with transported boulders (glacier or ice carried) without which many areas of those plains would have no stone for roads and buildings. The total quantity of glacial debris—clays, boulders, etc.—must be enormous, possibly thousands of cubic miles, but it is difficult to give any reliable estimate. There was also an Ice Age in the southern hemisphere, roughly 200,000,000 years ago. Evidence of glacial deposits tillite (or boulder beds) and of striated rock surfaces have been found in South Africa, Peninsular India, Australia, and South America. In India, in addition to continental ice sheets, there is clear evidence of icebergs having dropped facetted boulders in the sea which covered part of what is now the Punjab in Pakistan. The movement of the continental ice-sheets or glaciers in the Indian region, Gondwanaland of that late Palæozoic era, was northwards, and extensive deposits of the boulder bed (Talchir glacial boulder bed) occur in the Peninsula. These strata probably occupy an area of 50,000 square miles, either directly exposed or under the Damodar coal-bearing strata, and average, conservatively, a thickness of 50 feet. It is certain that these glacial deposits were more extensive when laid down. Assuming a spread of 100,000 square miles and a thickness of $52 \cdot 8$ feet (for convenience of calculation) the deposits represent 1,000 cubic miles in volume. Taking a cubic mile of detritus as about 10^{10} tons, all would weigh 10^{13} tons. If the contemporary glacial deposits—Dwyka of South Africa, basal Murree series of Australia, and the Itarar beds of South America are even the half of the Talchirs of India, the quantity will be 2,500 cubic miles (and roughly 25,000,000,000,000 tons), which is a smaller amount than may have been spread by the Pleistocene ice-sheets.

Sediment carried by Rivers

Various approximate measurements have been made of the silt and mud carried in suspension in some rivers. In the case of the Nile the estimate was 45,000,000 tons yearly carried into the delta from a catchment of somewhat more than a million square miles by 30 cubic miles of discharge water, which is 45 tons from each square mile of catchment. In the case of the Mississippi river the computation is 400,000,000 tons of silt into the Gulf of Mexico from a catchment of over 1,200,000 square miles with a discharge of roughly 150 cubic miles of water annually. This works out to about 320 tons per square mile of catchment. The average rainfall in the Mississippi drainage area is about 30 inches a year (so that only 25 per cent flows to the sea). The Yangtsi Kiang, with a catchment estimated at 650,000 square miles, is estimated to carry 320,000,000 tons of sediment to the East China Sea with a discharge of 165 cubic miles of water annually. This is nearly 500 tons from each square mile of catchment. F. W. Clarke (*The Data of Geochemistry*) has estimated that, for an area of 3,000,000 square miles in the United States of America, the amount of solid matter *removed in solution* annually is about 80 tons from each square mile. As will be seen later the proportion of solids removed in suspension as sediment is roughly three times the amount carried away in solution. On this basis the erosion of sediment from the 3,000,000 square miles of the U.S.A. is about 240 to 250 tons per square mile.

For comparative purposes the following data are of interest, as they cover drainage areas in different latitudes and include cold temperate and warm tropical areas. The Yukon River, draining a region between 66° and 60° north latitude, removes 90,000,000 tons as sediment and 20,000,000 tons of dissolved solids from a catchment of 320,000 square miles. The Columbia river, from a drainage area of 260,000 square miles between 54° and 47° north latitude, carries away 210 tons of silt in suspension and 70 tons of dissolved matter from each square mile of its catchment. This is less than the Yukon as regards silt but somewhat more in solution. The average for the rivers in the State of Washington is 200 tons of sediment in suspension from each square mile of the drainage area. In the State of Texas, between latitudes 33° to 29° N. and largely in the River Colorado basin (draining south into Matagorda Bay, Gulf of Mexico), the corresponding figure for removal of sediment in suspension is given as 8,000 tons per square mile. The Colorado river of Arizona, draining territory between 42° and 34° north latitude, carries 250 000,000 tons of sediment in suspension from an area of 240,000 square miles, which means 1,000 tons per square mile of its catchment.

If any reliance can be placed on the data for the silt carried in suspension in the cases given, particularly those for the Colorado rivers of Arizona and Texas respectively, which are remarkably high compared with the Mississippi or the Columbia of Washington, it would seem that the average for the rivers of the land may be greater than 264 tons per square mile of drainage area. This figure is difficult to assess since large areas of the land (reckoned at 57,000,000 square miles) are covered by ice, as for the Antarctic and Greenland and the islands between Baffin's Bay and the Beaufort Sea, and others that have an enclosed drainage, as the Sea of Aral, the Caspian, the Great Salt Lake, the Dead Sea, etc., and a considerable allowance is required. Allowing that these non-drainage areas cover 24,000,000 square miles there is a drainage area of 33,000,000 square miles from which at least 264 tons of sediment are removed per square mile and carried into the ocean. This amounts to roughly 9,000,000,000 tons a year, or 0·9 cubic miles of silt annually.

Solids carried in Solution by Rivers

While estimates for sediment in suspension are very unreliable as to the quantities actually carried into the oceans yearly, the data for computing the solids carried in solution in river waters are more satisfactory for rough approximations. Already it has been stated that the Yukon river, with a catchment of 320,000 square miles, carries 20,000,000 tons of solids in solution annually. This averages to nearly 63 tons from each square mile. In the case of the Columbia river, with a catchment of 260,000 square miles, the removal of solids in solution is at the rate of 70 tons per square mile. In the Colorado (Arizona) river, with a catchment of 240,000 square miles, the removal of solids in solution (11,000,000 tons in total) are calculated at 42 tons per square mile. In the case of the Mississippi river the solid matter in solution is estimated at 130,000,000 tons for a catchment of 1,250,000 square miles, or roughly 104 tons from each square mile. F. W. Clarke has made the estimate that the removal of solids from 3,000,000 square miles of the U.S.A. averages 80 tons from each square mile. He also gives the following figures for the land areas of the earth :—

Continent.	Area in millions of sq. miles.	Tons per Year per sq. mile.
North America	6	80
South America	4	50
Europe	3	100
Asia	7	85
Africa	8	45
Average for	28	70

This calculates to 1,960,000,000 tons, or roughly 0·19 or, say, 0·2 cubic miles a year.

Notwithstanding the fact that Australia contributes little to the oceans, either as flood or silt or dissolved solids, there is some contribution. Also the contribution from the Congo and Zambezi, catchments of nearly 2,000,000 square miles with discharges of upwards of ten times that of the Nile, would appear to suggest a higher figure for the African contribution. The same applies to the rivers of Asia, the Yangtsi Kiang and Hwang Ho, which have a total catchment of over a million square miles and total discharge of nearly 300 cubic miles. Finally, the figure for South America seems too low in view of the enormous discharge by the Amazon and La Plata rivers (nearly 4,000,000 square miles catchment and upwards of 1,000 cubic miles discharge). If the area of 28,000,000 square miles is safer than 33,000,000 square miles for the land areas which are supplying the dissolved matter, then it seems better to take an average from the European and American analyses and measurements, which are better known and cover cold temperate, warm temperate, and both humid and arid regions in climates. F. W. Clarke has shown, *The Data of Geochemistry*, 1924, p. 118, that Sir John Murray had taken an area of 40,000,000 square miles while his own data for salinity indicate nearer 100 tons per square mile (from American data) than 80 tons. On this basis the dissolved solids carried to the ocean would be nearer 4,000,000,000 tons, or 0·4 cubic miles a year. A figure between the two, 0·2 and 0·4, would be roughly the fairest average because it would allow for such large areas as do not contribute to the ocean and would permit of a higher general salinity for the rivers that do carry dissolved matter into the sèa. This gives a total of 2,680,000,000 tons, or 0·268 cubic miles, which again is about a third of the sediments carried in suspension.

The following analyses of the waters of the Amazon, the Mississippi, and the Nile show the nature of some river waters as well as changes (in the Mississippi) and other differences (in the White and Blue Nile) in the waters of such rivers in regard to the dissolved solids they carry :—

	(1.)	(2.)	(3.)	(4.)	(5.)	(6.)
CO_3	48·03	30·27	24·23	24·15	42·97	41·74
SO_4	9·35	19·69	32·74	2·26	0·25	5·62
Cl	0·83	11·05	3·15	6·94	4·38	2·19
Ca	20·77	20·25	15·04	14·69	9·78	18·38
Mg	7·27	4·66	4·37	1·40	3·00	4·66
Na and K	5·19	8·43	10·72	9·00	24·45	6·75
SiO_2	0·78	5·07	8·79	28·59	14·72	20·55
NO_3, Fe_2O_3	0·78	0·53	0·78	12·97	—	w. silica
	All the totals calculated to 100.					
Salinity—						
1 : 1,000,000	200	146	426	37	174	130
1 : 1,000	0·2	0·14	0·42	0·037	0·17	0·13

(1) Mississippi river, Minneapolis, Minnesota.
(2) Mississippi river, above Carrolton, Louisiana.
(3) Missouri river, near Kansas City, Missouri.
(4) Amazon river, at Obidos.
(5) White Nile, near Khartoum.
(6) Blue Nile, near Khartoum.

Public Relations Dept. S. Rhodesia.

PLATE XIV.—VICTORIA FALL, ZAMBESI.
Aerial view showing river dropping into chasm.

[*Facing page 80*

PLATE XV.—COAST EROSION ON THE NORFOLK SHORE.

PLATE XVI.—HEAVY SEAS POUND HASTINGS.
These high seas roll in during the November gales from the Atlantic and pound the coast of England.

The changes in the water of the Mississippi are seen in the following further analyses, which are also quoted from *The Data of Geochemistry* :—

	(1.)	(2.)	(3.)	(4.)	(5.)
CO_3	48·03	43·15	33·23	30·23	34·98
SO_2	9·35	12·55	21·74	20·50	15·37
Cl	0·83	2·21	3·79	4·10	6·21
NO_3	0·73	1·10	1·05	0·81	1·60
Ca	20·77	18·06	17·08	17·16	20·50
Mg	7·27	8·03	6·22	5·72	5·38
Na and K	5·19	5·52	8·15	9·61	8·33
SiO_2	7·78	9·03	8·54	11·44	7·05
Al_2O_3, etc.	0·05	0·35	0·20	0·43	0·13

All the totals calculated to 100.

Salinity—
| 1 : 1,000,000 | 200 | 203 | 269 | 202 | 166 |
| 1 : 1,000 | 0·2 | 0·2 | 0·27 | 0·2 | 0·17 |

(1) Mississippi river, Minneapolis, Minnesota.
(2) ,, ,, Quincey, Illinois.
(3) ,, ,, Chester, Illinois.
(4) ,, ,, Memphis, Tennessee.
(5) ,, ,, New Orleans, Louisiana.

These analyses show the upper Mississippi to be low in sulphates and chlorides. These come in from the Missouri. In Louisiana cyclic salt appears to be the source of the additional sodium and chlorine. Carbonates predominate in the Mississippi river waters.

The following are analyses of river waters draining arid regions which at times are hot countries :—

	(1.)	(2.)	(3.)	(4.)	(5.)	(6.)	(7.)
CO_3	31·91	33·68	7·34	37·55	2·65	0·96	9·31
SO_4	9·07	23·36	59·99	14·62	60·69	40·36	29·64
Cl	4·03	1·10	2·52	3·77	4·89	26·40	26·54
Ca	14·53	22·58	12·31	20·24	12·78	7·46	11·85
Mg	2·93	5·53	6·65	5·13	3·76	4·12	4·11
Na	10·80	5·12	9·84	9·57	14·50	20·64	17·03
K	2·72	1·66	0·34	0·60	0·28	—	—
SiO_2	23·50	6·49	0·94	8·19	0·45	0·06	0·34
R_2O_3	0·51	0·29	0·07	0·33	—	0·03	1·18
	100·00	100·00	100·00	100·00	100·00	100·00	100·00

Salinity—
| 1 in 1,000,000 | 37 | 137 | 1,571 | 148 | 2,134 | 6,670 | 1,182 |

(1) Cache la Poudre river, above North Fork, Colorado. This is a tributary of the South Platte, and the water is a typical mountain water pure, but high in carbonates and silica although with low salinity.
(2) Cache la Poudre river, sample from near Fort Collins, Colorado.
(3) Cache la Poudre river, sample 2 miles above Greeley, Colorado.
These last two analyses show a change to a hard sulphate water, largely the result of its use for irrigation purposes on the way.
(4) Arkansas river, Canon City, Colorado, and again downstream.
(5) Arkansas river, near Rockyford. The change from a carbonate to a sulphate water is again to be ascribed to irrigation *en route*.
(6) Chelif river, Algeria. Sample at low water from near Ksar Boghari, showing that the water has been in contact with salt and gypsum soils.
(7) Chelif river, downstream, near Orleansville, where it has become diluted. When the river is high at Ksar Boghari the salinity falls to 5,342, but the proportion of sodium to other salts in solution is greater.

Coastal Erosion by Currents and Waves

The chief ocean currents, such as the Gulf Stream, which sweeps the continental shelf of western Europe around the west coasts of Eire and Scotland, have a steady transporting action. Those along the west coasts of South America (Peru current), South Africa (Benguela current), and West Australia flow northwards. The equatorial currents flow westwards and in the crossing from Africa to South America one part of the south equatorial current swings southwards as the Brazil current and the other west-north-west across the mouth of the Amazon and carries away the silt brought down by this river. In the Arabian Sea, during the south-west monsoon a current develops from the Zanzibar coast to the coast of Malabar, but in crossing it a cold current is drawn up along the East African coast which kills coral growth. Corals grow abundantly in the warm waters of the Maldive and Laccadive islands, off the Indian coasts. Similarly, there is a southward surface current in the Bay of Bengal during the north-east monsoon which causes a cold stream of water to rise off the delta of the Ganges and which brings up deep sea material at the head of the bay. However, currents are affected by the tides, and sometimes accelerated and then reversed, as in the English Channel and the Straits of Dover. These tidal effects are further strengthened during storms and the waves then strike with great force and draw off with a powerful suction and cause great damage to coast and inland. Much depends on the nature of the cliff or shore, as well as the height of tide, the direction of the wind, and the force of the waves.

As an example of the action of currents which are also tidal a study of the erosion along the south coast of England is of interest. There is a pebble beach at Budleigh Salterton, in Devon, which checks the rough seas, but the pebbles are supplied from the cliffs of Bunter Pebble Beds (Triassic) as they are slowly carried along the coast into the sea. These pebbles reappear and supply the Chesil Bank, and again are carried along this natural breakwater until they pass on into deep water at Portland Bill. Here there is slow erosion largely controlled by the pebbles from the cliffs at Budleigh Salterton. Further east, from Beachy Head to Dungeness, in the Sussex–Kent coast, there are chalk cliffs which furnish flints but not in sufficient quantity to make a protective storm beach. It has been estimated that due to the violence of the sea waves the softness of the chalk, and the strong currents alongshore the cliffs are being eroded 3 to 4 yards deep year by year, and a quarter of a mile wide strip has been lost to the sea in the last 100 years. Along the east coast of England, where the rocks are also soft, say between Spurn Head and Flamborough Head, known as the Holderness coast, the erosion is severe. In this area, south of Bridlington, and particularly in the Hilderthorpe Cliff, a strip 100 yards wide has been washed into the sea during the past fifty years and of not less than 2 to 3 miles wide since the Romans were in Britain. A great deal of information on coast erosion was collected by the Royal Commission on Coast Erosion in 1909–1911 and was published in their valuable report.

The action of the sea is simple : a strong current sweeps away the material that can be moved ; a rise and fall of water level gives access to a wide shore, even to the cliff base ; wave action works on the strata ; each heave of the sea loosens material ; each withdrawal sucks out the free masses. When there is a gale of wind the waves are very much more powerful, and the effects of a single storm may erode more than the waves at high tide in fair weather over months of action. A strong off-shore wind by causing a set away from the coast draws up a current along the shore to the land, and may bring material from deep water to the shore. If there is a strong on-shore wind it tends to heap water on the shore, and this produces an undertow, dragging material into deep water. Under storm conditions the excavation by the action of a powerful undertow may produce an enormous amount of erosion, and even strip a beach of its sands and gravel. This heaping of water by a strong wind will cause shallowing on the windward side of a lake, and it has been suggested that under some such conditions of wind and shallow water the Israelites crossed out of Egypt. When the wind drops the waters return, but their power for erosion will depend on the force of the wind in the first case and the suddenness with which it dies away subsequently, and also by the open stretch of water to windward and the narrowness of the headwater. There is an almost " tidal " rise under some circumstances.

The influence of the tides, especially during high or spring tide, is well known. In the Bristol Channel the tide may rise over 40 feet along the South Wales coast or 20 feet higher than the neap tide. In the narrow Bay of Fundy the high tide may rise 70 feet above low tide at the head of the bay, or ten times higher than at the entrance. These phenomena become more accentuated in tidal rivers, when the high tide sweeps upstream on a flooded current flowing seawards. The tidal wave then races up as a wall of water, known as a " bore ", which may wash away obstacles along the shore and deposit them in the strong river current which carries them to the sea. The phenomenon known by the name *Seiche* is a swell or surging movement of the water in a large lake, and may be due to a combination of causes—action of a steady wind, tidal effect, and change in the atmospheric pressure. However, once the mass of water begins to oscillate it continues to do so like a pendulum, and with a definite period of to and fro movement until it slowly dies down. This kind of motion sometimes develops on the coast of Morocco in good weather and may make landing from boats difficult and dangerous. Storm waves resulting from cyclones, at the head of the Bay of Bengal, resemble gigantic " bores ", since they come on to a low coast and have a rise of 30 to 40 feet in an immense mass of water. The damage done by the Backergunge cyclone in and around the mouth of the Ganges was enormous. The storm drove shorewards at a time of high spring tide, and in less than half an hour had drowned the countryside to depths of 10 to 30 and 40 feet, and then the waters retreated quickly as the storm travelled inland. It is computed that, besides destruction of crops and cattle, over 100,000 people were drowned and a similar number died subsequently of cholera and disease.

Of the same nature as storm waves, but perhaps even more sudden owing to no storm warning or other cyclonic expectation, are the waves caused by earthquakes. After the Krakatoa earthquake a sea wave travelled across to Ceylon and was still 22 inches high there and 9 inches at Aden beyond the Arabian Sea. Charles Darwin has discussed the great earthquake of Conception. He wrote : " Shortly after the shock, a great wave was seen from the distance of three or four miles, approaching in the middle of the bay with a smooth outline ; but along the shore it tore up cottages and trees, as it swept onwards with irresistible force. At the head of the bay it broke in a fearful line of white breakers, which rushed up vertically to a height of 23 feet above the highest spring tides. . . . The first wave was followed by two others which in their retreat carried away a vast wrack of floating objects. In one part of the bay a ship was pitched high and dry on shore, was carried off, again driven on shore, and again carried off. . . ." He drew attention to the fact " . . . that whilst Talcahuano and Callow (in Lima), both situated at the head of large shallow bays, have suffered during every severe earthquake from great waves, Valparaiso, seated close to the edge of profoundly deep water, has never been overwhelmed, though so often shaken by the severest shocks."

Erosion caused by Man

The chief mischief produced directly by man is the erosion that follows deforestation in a region of high rainfall or one subject to torrential downpours. This is now well recognized, but it is doubtful if all the damage that has followed is yet under control. Erosion has been so great that in many areas the soil has practically been swept away and great floods have followed each other down the valleys. In some regions where gold deposits have been washed by hydraulic monitors large quantities of alluvial gravels have been carried downstream and resulted in floods which have covered good soils with sand or destroyed fishing of great value. These problems are generally fully attended to after considerable damage has already been done. Great floods have also resulted from the collapse of storage dams, such as the St. Francis Dam, in California ; but in these cases of failure an investigation is made and the inquiry produces information which reduces this kind of danger if care is exercised in subsequent works of the same kind.

CHAPTER V.—THE ACTION OF UNDERGROUND WATER

The chief source of underground water is that portion of the rainfall which sinks into the ground as infiltrating water or percolation. This water may be absorbed by the soil and taken up by plants or dried back into the atmosphere. Some of it may wet heavy clays and be tightly held in pore spaces of capillary dimensions, 0·0001 to 0·0002 mm. wide. The amount of water thus taken up may be very considerable since even a sandy loam has a pore space percentage exceeding 30, and a heavy clay may average a percentage higher than 50. Should the clay be so fine as to be colloidal then there may be an evolution of heat when such a clay is wet, as it will not only absorb water into its capillary openings but will also gelate. In both cases the clay will develop fluidity and, if present in large volumes, originate a stream of mud, often carrying rock debris on its surface as it flows across the country. It is not quite certain whether the heat given up is a mechanical phenomenon due to gelation of colloidal matter or is the result of hydration, like the slaking of quick lime. It is interesting to note that heat is also evolved when finely ground clinker (dry volcanic ash, burnt brick, Portland cement, etc.) is mixed with water and becomes fluid. It is now known, from experience in the making and use of " drilling mud ", that in addition to the viscosity that these liquid muds possess they may also have the characteristic of behaving like a jelly. This behaviour, by which a mud may *set* like a jelly when undisturbed, is known as thixotrophy. It means that a mixture of water and colloidal clay has been converted into something more intimate than an emulsion or permanently suspended clay in water.

Percolation and Absorption

The term absorption seems suitable for the water that may be taken up by clays in their pore spaces and by plants into their tissues and vessels, while the term adsorption can refer to the gelation of colloids and the slaking of lime or setting of cement. Although such water is grouped with the percentage of rainfall disposal called percolation, it is hardly percolation in the sense that it passes underground ; and, generally, it will be given back to the air when subject to strong heating by the sun's rays. Mention has already been made of percolation and infiltration on p. 57, and it was there discussed how the size of the particles, or really the friction of passage through the pore spaces, affected the rate of percolation. Fine material, with capillary openings, behaves like blotting paper or a sponge, while coarse material allows water to flow more or less easily through it.

Intermediate between an impervious wet material and a highly porous stratum there may be medium-sized pore space cavities which offer some resistance to the flow but do not prevent it. This kind of movement of underground water may be termed seepage. The water, entering a well through the pore spaces of a water-bearing sand, does so by infiltration, if the well is a good source of supply. It may be remarkable because of large fissures admitting the water, but it may be a poor well if it fills slowly by seepage.

The rate of infiltration into a well or trench or other excavation is of the greatest importance, since it may prove to be a grand source of water supply, or it might become a costly item to deal with in tunnelling or shaft-sinking operations. It is not possible to predict the yield of water from fissured rocks, such as granite or limestone, as much depends on the geological mode of occurrence of the rocks and the configuration of the ground, but in porous strata, such as sandstones and pebble beds, something can usually be expected if the geological structure of the area is known. In his book *The Nation's Water Supply* (1936), R. C. S. Walters, discussing " Geology in Relation to Water-Supply " and " The Underground Resources ", draws distinctions between *impermeable strata*—igneous rocks, metamorphic rocks, Cambrian, Ordovician, Silurian, Old Red Sandstones, and Devonian, Culm Measures, part of the Carboniferous, Lias, Oxford, Kimmeridge, and Gault clays, and the *permeable* strata—Carboniferous, Coal Measures, Trias, Oolites. Lower greensand, chalk, and drift (boulder beds) for upland and underground resources in England. The impermeable rocks would prove most suitable for sites for reservoirs and dams, while the permeable strata are those in which wells and boring would be sunk to tap percolating water. Mention is made of leakage through fissures and fault planes in the impervious rocks, which led to the collapse of a dam and disaster. It is also noted that the water in the chalk usually occurs in the joint and bedding planes and only a small proportion in the rock itself. It is for this reason that one well at a distance may seriously affect another well in the same stratum.

With regard to water-divining in England, Walters has made the following interesting comment : " . . . As far back as the time of William Smith, the ' Father of English Geology ', it is related how he walked with a diviner on the Mendips a century ago and surreptitiously dropped pebbles at every place where water was ' found '. On returning over the same ground, none of the previous places were pointed out, but other places were indicated, and Smith advised that ' as the water had changed its situation at all points, it would be imprudent to spend money following it '." Since then many experiments have been carried out in England and elsewhere to test the ability of water-diviners, and it is to be said that in the majority of such examinations water-divining was useless and proved quite incapable of " finding " water that was already known to occur. The results of such investigations have decided most engineers and contractors to secure trained geological advice, since the geologist must not only unravel the structure of the rocks but must also know their nature and texture, and make his report on lines which are intelligible to any educated person. It is true that a large quantity of water has been " found " in a well in hard impervious rocks (in a plane of fracture), and that another well has failed to get water in porous sandstone (because the sandstone was argillaceous at the locality), but good local knowledge (by close mapping) might have elucidated both positions previous to making the wells.

Unconsolidated sands and gravels may have pore space volumes of 25 per cent and prove, as in the case of sand dunes, either inland or on the coast, very useful sources of water supply. Where such sands have been buried to a depth of 200 to 400 feet, and thus are subject to the weight of superincumbent beds, the pore space volume will be reduced perhaps 12 to 15 per cent if the material is coarse. In sands the pore space may be reduced to 10 per cent if the material is medium grained, and the ease of percolation will also be reduced, so that, after pumping, a well may not fill again quickly. At greater depths, when the coarse sands and gravels have been consolidated into sandstones, the porosity may be still further reduced by the pressure packing the grains very tightly, but this may not affect the easy passage of the water through the pore spaces to any marked extent unless the pores become choked with clay or deposits of silica, iron oxide, limestone, etc. It is usually possible to ascertain whether such cementing material is present, but for important supplies the facts are very quickly determined by boring (a core boring). The boring may become the means of drawing off the water subsequently. The geology of the area must eventually be elucidated if a well is very productive, since other wells will follow and interference be caused if the replenishment of the water-bearing stratum is not adequate.

In his *Treatise on Metamorphism* (1904) Van Hise has stated (p. 142) : " . . . the artesian water adjacent to Lake Michigan, at Chicago, in the early wells, before they became so numerous as to interfere when allowed to flow, had a head of 30 metres above the surface, and the feeding area is only about 80 metres above Chicago ; yet the water travelled underground from 150 to 250 kilometres. The resistance causing the loss of head of 50 metres is to be distributed through this distance ; therefore the friction per metre must have approached an infinitesimal amount. . . . In all such instances the average movement is exceedingly slow . . . the moment the speed of movement becomes appreciable the resistance promptly runs up. . . ." And several observers have written on the freshwater springs of the Bahrein Islands, in the Persian Gulf. G. E. Pilgrim, in 1908, wrote of these as follows : " It is certain that these great supplies of water are of artesian origin and are derived from the elevated country in the interior of Arabia and the highlands of the Nejd, where it is likely that the rainfall is fairly large. . . ." At an earlier date, 1900, S. M. Zwemer had written : " Arabia has no rivers and none of its mountain streams (some of which are perennial) reach the sea coast. At least they do not arrive there by the *overland* route, for it is a well-established fact that the many freshwater springs found in the Bahrein archipelago have their origin in the uplands of Arabia." These uplands about Riyadih are from 300 to 400 kilometres to the south-west.

The importance of the artesian water in Queensland has long been understood. Already in 1904 upwards of 400,000,000 gallons of water a day were obtained from 1,000 wells (600 were flowing wells) from depths of 2,000 feet. At Blackall, in central Queensland, a well which had tapped water at 1,645 feet (or 775 feet below sea-level) was under such artesian conditions that the water flowed out of the top at nearly 300,000 gallons a day. These artesian waters are believed to draw their supplies from the rain which falls abundantly on the mountains bordering the coast—the Great Dividing Range. At Thargomindah, in south-west Queensland, the pressure of the artesian water was enough to drive a turbine coupled to an electric-lighting set, and at another power is obtained through a pelton wheel fixed over an artesian well. However, the water was of greater use for watering cattle raised on the western downs when crossing the deserts of Lake Eyre on their way to the railway at Hergott, in South Australia. In this region, along the south-western and southern side of the almost dried-up bed of Lake Eyre, are the famous Mound Springs. They do not derive their water from local rainfall (which is practically nil) but from some deep-seated source as they are hot springs and contain abundant minerals in solution. It is the deposit from the waters that has made the " mounds " at the springs. A thermal spring at Clifton, New South Wales, tapped by a bore-hole at a depth of 1,638 feet, has a temperature of nearly 140° F.

Weathering of Rocks

Rain-water carries with it carbonic acid and other constituents from the atmosphere, which react on minerals as the water percolates into the ground. Much of this attack is to form hydrated compounds, but the consequences are that the percolating waters remove in solution several constituents. Observations show that practically all known rock-forming minerals are liable to attack by percolating rain-water. Where the exposed rock is on elevated ground, but the slopes are low, the removal of soluble matter is greater than on steep slopes. In the formation of in situ or primary laterite the rock is gradually rendered porous and subjected to drenching rain which soaks the pores and then slowly drains away. This alternation occurs once a year in tropical regions. In process of time the result is a crust or rind of porous rock largely consisting of the hydrated oxides of aluminium and ferric iron, as much as 30 to 40 feet thick, lying nearly horizontally on kaolinized rock which passes down into unaltered rock. If the original rock is granite with quartz, the results of the weathering are similar in appearance but the alumina and iron oxides are largely in the form of hydrated silicates, clayey or lithomarge products. If the rock is doleritic basalt the resulting products are hydroxides of alumina and ferric oxide. In the granite weathering the insoluble quartz is left, but in the case of dolerite, where there is no free quartz, the silicates are decomposed and the silica removed in solution. Laterite formation, both from granite yielding the lithomargic product and from the dolerite yielding a bauxitic product, appears to be restricted to tropical regions subject to monsoon (alternate dry and wet periods) climates (see " Buchanan's Laterite of Malabar and Kanara ", by Cyril S. Fox, *Rec. Geol. Surv. India*, vol. lxix, pt. 4, 1936).

The alterations mentioned in the previous paragraph are seen in the analyses given below :—

	Alteration of Basalt.			Alteration of Khondalite.		
	(1.)	(2.)	(3.)	(4.)	(5.)	(6.)
Silica	48·62	37·31	1·44	60·08	45·14	16·23
Titania	0·88	3·33	6·32	0·65	trace	0·93
Alumina	14·12	27·85	62·32	12·38	36·69	26·82
Ferric ox.	2·29	17·36	2·65	3·28	2·24	41·69
Ferrous ox.	12·40	—	—	4·20	—	—
Lime	9·49	—	—	9·43	1·07	—
Magnesia	5·29	0·76	0·38	1·95	0·15	—
Alkalies	3·55	—	—	5·65	—	—
Comb. water	2·28	13·40	26·70	1·80	11·65	14·20

(1) Basaltic lava from Kolhapur State (Deccan trap).
(2) Kaolinized basalt from Kolhapur State (Deccan trap).
(3) Aluminous laterite resulting from decomposition.
(4) Khondalite (fresh rock, analysis by M. S. Krishnan).
(5) Altered Kalahandi Khondalite (analysis by M. S. Krishnan).
(6) Laterite, Korlapat, Kalahandi (analysis by M. S. Krishnan).

Analyses of the lithomarge found in association with these laterites range from 42 to 25 per cent silica, 40 to 28 per cent alumina, 0·5 to 25 per cent ferric oxide, and 8·0 to 14·0 per cent combined water. These analyses confirm the processes mentioned above, and have been confirmed by others of samples from British Guiana, Malay States, etc. They show the leaching of the alkalies, lime, magnesia, and also of silica (for aluminous laterite) and hydration of silicates (for forming lithomarge or kaolin).

In the process of lateritization, discussed above, the accompanying development of lithomarge is similar to that of true kaolin. Lithomarge is a loose term and includes kaolin, but kaolin or kaolinization occurs under the mantle of ferruginous laterite. Kaolinization occurs in other ways and has been ascribed purely to the attack of waters carrying carbonic acid on feldspars and taking place to appreciable depths from the surface with no relation to laterite. In both these examples the rock-forming minerals, usually the feldspars, are decomposed and permit the water to carry away soluble components and to form hydroxides with those that remain as insolubles, except, of course, quartz or so-called free silica. In the case of other rocks, the mode of attack may be different, and in the case of limestone it may be a direct solution of the carbonate of lime, or a replacement action if the water is already carrying another carbonate, say magnesia or ferric iron. In the latter case the water will exchange the ferric oxide for calcium carbonate and so leave a deposit of iron ore in a limestone. The reactions above described are all more or less at the surface or within relatively shallow depths, but much depends on the configuration of the surface, on the structure of the rocks, and their nature as to how superficial or how deep kaolinization may be carried.

The percolating waters involved in the processes of rock weathering frequently emerge as springs and join the run-off rain-water in the rivers, and, of course, bring the dissolved material with them in most cases. Some of the percolating water continues down to replenish underground movement. A computation shows that for every average cubic mile of river water, run-off rainfall, weighing roughly 4,090,000,000 tons, there is contained the following amount, in tons, of various constituents leached from the rocks :—

326,710	tons of	calcium carbonate ($CaCO_3$)
112,870	,,	magnesium carbonate ($MgCO_3$)
2,913	,,	calcium phosphate ($Ca_3P_2O_8$)
34,361	,,	calcium sulphate ($CaSO_4$)
31,805	,,	sodium sulphate (Na_2SO_4)
20,358	,,	potassium sulphate (K_2SO_4)
26,800	,,	sodium nitrate ($NaNO_3$)
16,657	,,	sodium chloride (NaCl)
2,462	,,	lithium chloride (LiCl)
1,030	,,	ammonium chloride (NH_4Cl)
74,577	,,	silica (colloidal — SiO_2)
13,006	,,	ferric oxide (hydrated — Fe_2O_3)
14,315	,,	alumina (hydrated — Al_2O_3)
5,703	,,	manganese oxide (Mn_2O_3)
79,020	,,	organic matter, making a total of
762,587	,,	dissolved matter per cubic mile of river water (see *Geology: Processes and Their Results*, vol. i, 1904, p. 107, by Chamberlin and Salisbury).

In contrast with the above, one cubic mile of sea water carries in solution the following components by weight :—

117,474,000	tons of	sodium chloride (NaCl)
16,428,000	,,	magnesium chloride (MgCl)
7,154,000	,,	magnesium sulphate ($MgSO_4$)
5,437,000	,,	calcium sulphate ($CaSO_4$)
3,727,000	,,	potassium sulphate (K_2SO_4)
328,000	,,	magnesium bromide ($MgBr_2$)
521,000	,,	calcium carbonate ($CaCO_3$), or a total of
151,025,000	,,	dissolved salts per cubic mile of sea-water (see *Challenger Reports: Physics and Chemistry*, vol. i, p. 204).

The above data become of greater interest if compared with the solubilities of different substances in water at different temperatures, as shown in the list below :—

Substance or Solute	Amounts in grams per 100 grams of Water.		
	0° C.	20° C.	100° C.
Sodium chloride	35·7	36·0	39·8
Potassium nitrate	13·3	31·2	247·0
Barium chloride	30·0	35·7	58·8
Calcium carbonate	0·0018	—	0·0018
Copper sulphate	15·5	22·0	73·5
Silver nitrate	121·9	227·3	1111·0

It is thus seen that common salt is about 20,000 times more soluble than limestone, which is among the most soluble of the rocks exposed to weathering and the solvent action of percolating water.

The analyses of some river waters has already been given on a previous page, but it is of interest to notice the analyses of the Yukon river on the borders of the Arctic and of the Thames in England, and the pure waters of the Dee, near Aberdeen, in Scotland :—

	(1.)	(2.)	(3.)
CO_3	44·03	39·53	23·35
SO_4	13·99	14·72	15·70
Cl	0·01	4·57	17·08
Ca	24·38	28·57	17·22
Mg	5·23	1·82	2·98
Na and K	4·71	4·83	13·60
SiO_2	7·52	2·36	6·41
R_2O_3	—	3·60	3·66
	100·00	100·0	100·00
Salinity in a million	122	266	31

And a general average of the river and lake and oceanic waters is shown below for comparison :—

	Lakes and Rivers of the World.	Oceanic Waters.
CO_3	35·15	0·213
SO_4	12·14	7·914
Cl	5·68	55·185
NO_3	0·90	—
Ca	20·39	1·244
Mg	3·41	3·896
Na	5·79	30·260
K	2·12	1·109
Br	—	0·179
R_2O_3	2·75	—
SiO_2	11·67	—
	100·00	100·00

If the salinity of the oceanic waters average 3·5 per cent, or 35,000 in parts per million, the average of river and lake waters are probably $\frac{1}{200}$th, or about 175 parts per million. All these analyses have been quoted from F. W. Clarke's great compilation, *The Data of Geochemistry*, 1924.

It is generally understood that the zone of weathering extends downwards to as far as air is carried. The chief action of percolating water in the zone of weathering is oxidation accompanied by solution. The waters thus loaded with dissolved matter, as indicated above, may return to the surface and to the salinity of the river waters. Some of the analyses which have been quoted, as in the case of the Mississippi and also the Cache la Poudre rivers, show how the composition of the dissolved salts may be altered by tributaries and by dilution. Eventually the main effect is that of a water holding the more soluble components, just as the oceanic waters hold sodium chloride in far greater proportion than calcium carbonate. Such details are well seen in the study of rock-salt deposits, as those in the Magdeburg-Halberstadt region, where the Stassfurt beds show gypsum at base followed by salt, polyhalite, kieserite, and carnallite at the top, as succeeding deposits as evaporation progressed and the less and less soluble components were precipitated.

Underground Rivers and Springs

The following notice appeared in the *Daily Telegraph* of 6th January, 1949 : " Boy fell into hidden stream," and stated that a child of three disappeared into a hole in the grounds of Ven House, near Milborne Port, Somerset, while playing there. It was discovered that the body was in an unknown underground stream and had been carried 10 yards along its hidden channel. This is a case of recent occurrence in England, so that it is evident that this kind of action is not restricted to spectacular examples as the Mammoth Caves of Kentucky. Indeed, underground rivers and caves are common. There are the Carlsbad Caverns of New Mexico, the Caverns of Luray, in Virginia, in the United States, those of Jenolan in New South Wales, Australia, and other countries. There are the tourist cave of Adelsberg, Postumia, and the grotto of St. Canziano, into which the rivers Puika and Timavo flow, respectively, in the region north-east of Trieste, and so disappear in the limestone rocks of Gorizia. It is thought that these rivers eventually discharge into the Gulf of Panzano at the head of the Adriatic in the sea bed. In England there are the well-known Cheddar Caves, in Somerset, and those of Clapham, in Yorkshire. In all these cases the run-off flow has sunk into cracks in limestones and enlarged them below ground into extensive chambers and caverns, and continued downwards to the level of standing water and there supplied the main underground storage, or ultimately flow into the sea. Mention has been made of the great springs in the sea along the Italian and Grecian coasts—at Syracuse (Sicily), Taranto harbour, Gulf of Spezia, etc., in Italy, and the Gulf of Corinth (Delphi) and other places which appear to be fed from limestone hills and highlands (karstenlands) through which the currents and streams flow with great volumes

THE ACTION OF UNDERGROUND WATER 93

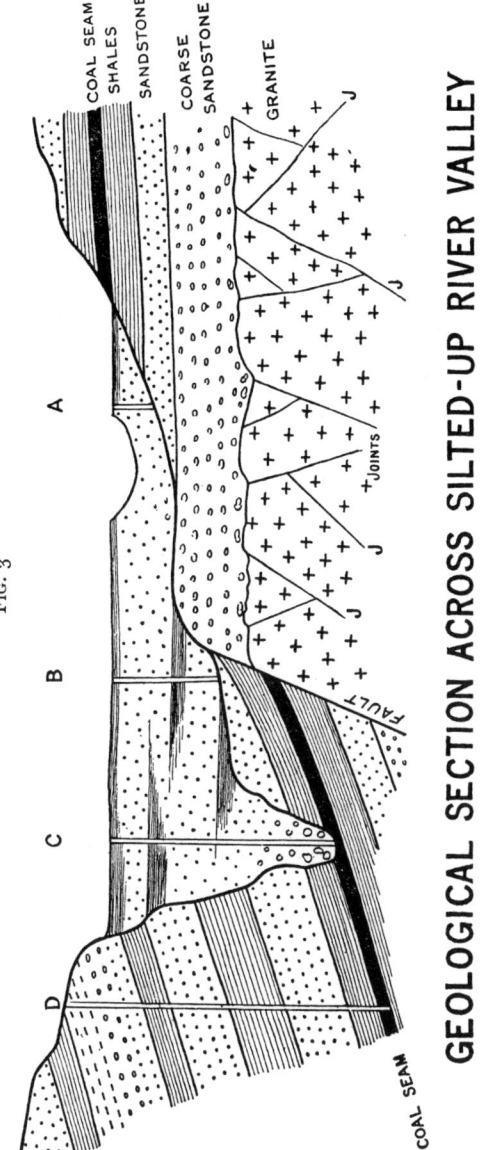

Fig. 3

GEOLOGICAL SECTION ACROSS SILTED-UP RIVER VALLEY

A. Shallow well into sandy alluvium near stream bed.
B. Deeper well after boring proves bed sands.
C. Deep boring into gravel filling an old bed.
D. Shaft to meet stratum which taps the old bed and up which a drivage will go.

of water. These underground rivers work both by the solvent action of water and by abrasion since they carry detrital material, sand and gravel, and even boulders from the surface through sink holes (swallow holes) into their underground channels.

In the case of the great Australian artesian basin of south-west Queensland, the porous or water-bearing beds are Jurassic strata which are overlaid by impervious Cretaceous beds. The rainfall is carried in from the high lands of eastern Queensland and sinks down and percolates more than 300 to 400 miles south-westwards. The deepest artesian well is reported to be 7,000 feet deep, and such waters are under great pressure, partly from hydrostatic head, but perhaps by the direct weight of the overlying strata. These questions are of considerable interest as they would establish a definite example of how the water in the pore spaces of a porous rock are naturally squeezed out as the strata are more and more deeply buried under newer formations. In the Australian artesian basin there would appear to be evidence that the deep borings are not exhausting the artesian supply, so that the percolation must be in progress to replenish whatever is being drawn off. In the case of the artesian water around Chicago the numerous wells long ago interfered with each other while allowing an easy escape for the water under pressure, and as a result the supply is not under as great pressure as originally. This percolation and infiltration, by being tapped and put in quicker movement, may increase the rate of removal of solid matter in solution. In process of time there may be some subsidence by such removal, but the danger is probably more in the case of active underground rivers than in percolation, which can increase the pore spaces and improve the rate of infiltration and thus yield a larger amount of water from a given well.

That changes do occur in the flow of underground waters is seen in various wells and springs. The most historic example is the Pool of Siloam, which overflowed periodically and the waters of which were regarded as curative at such times. Ebbing and flowing wells are readily explained in coastal positions since they may be influenced by the rise and fall of the tides even though the wells contain fresh water. Charles Darwin has stated that such wells are common in some of the low islands of the West Indies, but in these cases if the fresh water flows outwards through coral with a very open texture there is likely to be some contamination with sea water, as the fresh water has not enough head to wash out the salt water. A spring or well of this kind occurs at Newton Nottage, near Bridgend, in Glamorgan, which draws its water from the Carboniferous limestone and a conglomerate. The well is 500 yards from the shore at high water and 8 feet above the sea-level, but it rises and falls with a 3-hour lag on the tides, and the water is slightly brackish. The Pool of Siloam, like other intermittent wells—the Fountain of Miracles of Galero, in the Ligurian Alps, etc.—

THE ACTION OF UNDERGROUND WATER

appears to connect with underground caverns which fill slowly and automatically empty themselves by moving a boulder, ball-value fashion, or by compressing air in the cavern. The curious flowing of the Croydon Bourne appears to be affected by similar action. And this appears to be the best explanation of the filling of some of the Norfolk meres which draw their water from the chalk. No sound appears to confirm these floods of water, but blowing wells are known when the sound of escaping air is heard, due to expulsion of air when the ground water-level rises (and also when there is a sharp fall in the barometer). It is believed that the " oracle at Delphi " *spoke* by some similar compression of the air in a cavern (which is near and probably connected with the great freshwater springs which emerge in the Gulf of Corinth).

To give some idea of the size of the cavities which are produced by solution by rivers flowing underground, it is well to remember that the grotto of Postumia (Adelsberg) has a small tramway for taking tourists underground for a mile or more, after which there is a further walk (and still further for exploration). The Mammoth Caves in Kentucky, which have long been a National Park, are in the St. Louis limestone under the Chester sandstone (both Lower Carboniferous in age). The caverns have been made since Miocene (or Pliocene) times. The limestones cover an area of 8,000 square miles and are traversed by fissures along which the water has acted. The main cavern is from 40 feet minimum to 300 feet maximum in width, and from 35 to 120 feet high, but the galleries and chambers and domes extend in all directions for 150 miles There are streams and pools of considerable size and a full exploration took several years to complete, in so far as the more accessible places are concerned, but it is not yet known how deep the waters go, nor how widely the joints and bedding planes have been subject to solution by the flowing waters. In the case of the Carlsbad Caverns National Park in New Mexico, the enclosing rocks are salt, gypsum, and limestone and the Big Room is 800 yards long, 250 feet high, and as much as 600 feet wide at one place. The material removed from the Big Chamber alone must exceed 8,000,000 tons of relatively soluble material, while that from the main cavern of the Mammoth Cave, even for a mile, will amount to nearly 2,500,000 tons of much less soluble matter. The total may be greater than 250,000,000 tons of limestone, say, in 12,500,000 years (or roughly only 20 tons a year, which is very small).

Thermal Springs and Volcanoes

When mentioning the Mound Springs of Australia it was stated that some were thermal waters containing a large amount of mineral matter (which deposited the mounds). It has been thought that these waters, because of a temperature of 140° F. or so, are of deeper origin than those of the great artesian basin of Queensland; but the argument is doubtful since these hotter waters of South Australia come up along fissures (a line of faulting ?) and so would not lose as much heat as water percolating through the pores of a porous stratum. The artesian wells along the coastal tract of Western Australia, near Perth, even when tapped at depths of 1,800 feet do not show temperatures of 100° F., and so are regarded as meteoric in origin. It is commonly accepted that the rate of increase in depth into the earth's crust is 100 feet per 1° C., or roughly 66 feet per 1° F., but this average is for a wide range of rocks and depths. The oldest rocks, Archæan gneisses and Precambrian strata, show greater depths per 1° C. and some Mesozoic and Kainozoic strata show higher temperature gradients than the average. In the case of very deep borings, say that by the Continental Oil Company, 4 miles west of Wasco, San Joaquin Valley, California, which was carried to a depth of 15,000 feet in Miocene sands, the temperature readings were 196° F. at 6,000 feet, 268° F. at 15,000 feet, which suggests only 1° F. for 125 feet (or half the average rate). In matters of this kind no opinion can be advanced with certainty unless the geological structure of the region is known and the water resources have also been carefully studied.

Whenever sediments are laid down by water the pore spaces between the grains are full of water. The proportion may vary from 60 to 30 per cent of the volume of the deposit, larger in fine materials, clays, etc., and smaller in coarse debris, gravels, and pebbles, etc. But when weight is applied to the bed, by further sedimentation or by dynamic action, the sediment will be compressed and the water squeezed out. In the case of clays, the material is transformed into shale and its porosity greatly reduced, to less than 5 per cent, and all the interstitial water is eliminated. In the case of pebble beds, there is much greater resistance to pressure and although the packing may become tighter, the rock still has a high porosity above 12 per cent—and remains porous. The pressure exerted by the weight of superincumbent beds is itself able to generate heat in the stratum which is compressed. By compressing a firm brick clay sufficiently the temperature of the material is raised, and experimental work has shown an increase from 18° C. to 40° C., which is equivalent to a burial of nearly 2,000 feet. N. L. Bowen and M. Aurousseau (*Bull. Geol. Soc. America*, vol. 34, 1925, p. 431) have discussed "The Melting of Sedimentary Rocks in Drill Holes". They recorded the melting of shale and sandstone under the pressure of a 4 in. steel casing at a depth of 4,300 feet under a load of 20 tons. The drill was rotated at about 20 feet per minute and stuck after penetrating the argillaceous stratum to a depth of 3 feet. Thus, rock pressure may, as a result of some movement, shearing particularly, raise the temperature locally at great depths and heat up any escaping water.

PLATE XVII.—FLOODED VALLEY OF THE RIVER SEVERN, ENGLAND.

PLATE XVIII.—FLOODED RIVER COVERS WATERFALL.

This view of the Columbia River in flood below Wenatchee, Washington, shows rapids where the Rock Islands falls can be seen under normal conditions. Due to the loss of " head ", 51 feet to less than 10 feet, the hydro-electric plant could not produce enough electricity for the demand. This raises the question whether, in such cases, the generating plant should not be lower downstream. The intake would still be above the falls but the discharge from the turbines would be below the " boil " of waters under the falls.

[*Facing page 96*

PLATE XIX.—GRAND COULEE DAM, COLUMBIA RIVER, WASHINGTON STATE.

Here is the Grand Coulee Dam, largest concrete dam in the world and kingpin for the Bureau of Reclamation's millio .-acre Columbia Basin Irrigation Project. This multiple-purpose dam raises the level of the Columbia River to facilitate irrigation pumping planned for the project. Power generated by the dam will be used to operate the world's mightiest irrigation pumps that will start Columbia River water on its way to a million-acre expanse of farming land. The world's biggest hydro-electric generators are operated at the dam. Its ultimate power installation will be the greatest in the world. The West Powerhouse, farthest from the camera, already contains the greatest installed capacity of any single powerhouse in the world.

THE ACTION OF UNDERGROUND WATER

There have been many contributions relating to thermal springs, e.g. "The Temperature of Hot Springs and their Sources of Heat" (see *Journal of Geology*, vol. xxxii, 1924). A. L. Day, E. T. Allen, L. H. Adams, and C. E. Van Ostrand considered that the heat is largely produced by chemical processes and is volcanic, that radio-activity is not a great factor, and that the water is largely of surface (meteoric) origin in circulation, though some magmatic (juvenile) water is probable. Geologists consider that all the water is meteoric (vadose) if the rocks are of sedimentary origin. In the Ridgeway boring (for oil in England), which met the Carboniferous limestone at a depth of 2,815 feet (2,306 feet below sea-level) and continued in it to a total depth of 2,898 feet (from the surface), a hot water horizon was tapped. The water flowed out of the top of the boring at 100,000 gallons a day and at a temperature of 120° F. This, if the air temperature be assumed at 68° F., works out to 100 feet per 1° C. descent. The water contained the following constituents in solution in parts per 100,000 :—

Calcium carbonate	13·75	parts per 100,000
Calcium sulphate	173·79	,, ,,
Strontium carbonate	1·50	,, ,,
Magnesium sulphate	15·35	,, ,,
Magnesium chloride	34·92	,, ,,
Sodium chloride	162·55	,, ,,
Lithium chloride	1·20	,, ,,
	403·06	,, ,,

Salinity in parts per million is 4,030·6.

The high proportion of sodium chloride (as shown in the analysis) suggests that there is an appreciable amount of residual sea water present, presumably from the marine beds.

The following analyses of other mineral, but not thermal waters (from *The Data of Geochemistry*, 1924, p. 184 onwards) are of interest both in showing evidences of sea water and other types :—

	(1.)	(2.)	(3.)	(4.)	(5.)	(6.)
Cl	61·65	52·74	62·31	61·38	0·48	4·01
Br	0·29	—	0·53	0·26	—	—
Iodine	trace	—	0·01	trace	—	—
SO$_4$	0·07	8·36	0·03	0·02	66·28	4·26
CO$_3$	—	1·88	0·27	nil	0·60	47·45
Na	31·57	30·15	18·35	24·50	30·46	40·09
K	trace	—	1·55	1·97	1·08	0·38
Ca	4·85	4·74	13·86	9·56	0·67	0·30
Mg	1·52	2·09	2·53	0·94	0·41	0·12
SiO$_2$, etc.	0·05	0·04	0·29	0·06	0·02	3·39
	100·00	100·00	100·00	100·00	100·00	100·00
Salinity reckoned at 1 in a million parts	178,900	15,905	309,175	263,640	74,733	1,668

(1), (3), and (4) are clearly natural brines and classed as chloride waters with (2), while (5) and (6) are, respectively, sulphate and carbonate waters. The particulars are :—

(1) Artesian well, 1,260 feet deep, Abilene, Kansas.
(2) Deep well, 1,505 feet depth, Brunswick, Missouri.
(3) Brine well, 2,667 feet deep, Conneautsville, Pennsylvania.
(4) People's natural gas well, 6,300 feet deep, 8 miles south-west of Imperial, Washington Co., Pennsylvania.
(5) Abilena well, 130 feet deep, 14 miles north-west of Abilene.
(6) Artesian well, 386 feet deep, La Junta, Colorado.

As the flow is not stated it is not possible to estimate the quantity of solids removed in solution, but the salinity of (3) is nine times that of normal sea water.

Mineral springs occur in almost every country of any size. Although hot springs are not so common they are nevertheless not rare, except those which discharge their waters at nearly boiling point. In such cases, hot springs are usually found along lines of great faults, or on the edge of important igneous intrusions, or again in areas which may be regarded as still volcanic. Armand Gautier has pointed out that meteoric waters (from the surface or vadose) affect springs in characteristic ways—causing fluctuations of composition, concentration, and rate of flow, depending on the rainfall. Such waters also contain carbonates of lime and magnesia, chlorides and sulphates. Virginal or juvenile waters as a rule have a steady flow and constant composition, carrying sodium bicarbonate, alkaline silicates, heavy metals, etc., as chief constituents, with chlorides or sulphates as accessories, and practically no carbonates of the alkaline earths. These generalizations may be decided at once by the relative steadiness of hot springs and geysers (which are intermittent hot springs). That most amazing region, the Yellowstone National Park, in north-western Wyoming, contains 100 geysers and upwards of 4,000 hot springs, including a steam jet in the Yellowstone Grand Canyon, 1,000 feet below the rim of the main plateau. On the other hand, the hot springs of volcanoes carry strongly acid waters; but even this criterion may prove unreliable, since some mine waters may contain acid (sulphuric acid) derived from the weathering of iron pyrites. However, in most cases a geological study will usually give data to support a meteoric origin or give reasons in favour of the presence of juvenile waters. In all these cases the waters generally carry large quantities of dissolved matter and are consequently actively engaged in erosion by solution.

The following analyses, quoted from *The Data of Geochemistry*, 1924, p. 188 onwards, show the waters of hot springs, geysers, and volcanoes, but they are exceedingly variable and can hardly be used for comparison except in a very general way :—

THE ACTION OF UNDERGROUND WATER

	(1.)	(2.)	(3.)	(4.)	(5.)
Cl	58·79	1·27	31·64	13·52	37·52
Br	trace	trace	0·25	—	—
SO_4	0·94	3·39	1·30	9·01	4·96
CO_3	0·61	41·47	8·78	10·16	—
Na	30·38	2·38	26·42	19·71	24·22
K	3·76	0·80	1·93	1·88	0·36
Ca	4·90	23·54	0·11	—	2·56
Mg	0·40	2·56	0·04	0·08	0·19
Fe and Al	—	0·27	0·12	—	0·35
SiO_2	0·20	22·85	27·58	45·04	29·81
	100·00 [1]	100·00 [2]	100·00 [3]	100·00	100·00
Salinity in parts per million	23,309	284	1,388	1,131	2,735

(1) Utah Hot Springs, 8 miles north of Ogden, Utah. This water is at a temperature of 55° C. (131° F.) and is a brine.
(2) Big Iron Hot Spring, Arkansas.
[1] The water contains also NO_3 0·23 per cent, PO_4 0·03 per cent, BO_2 0·64 per cent, NH_4 0·03 per cent, also Li, Ba, Sr.
(3) Old Faithful Geyser, Yellowstone National Park, Wyoming. It erupts every 65 minutes for 4½ minutes, to a height of 100 feet or so. Others, like the Giant Geyser, in the same region, erupt to 60 feet plus a squirt to 250 feet.
[2] The water contains also AsO_4 0·24 per cent, B_4O_7 1·19 per cent, Li 0·4 per cent and Mn.
(4) Great Geyser of Iceland, area 30 miles north-west of Hekla, issues at a temperature of 76° to 89° C. (168° to 188° F.), from a basin 60 feet across and 4 feet deep, to a height of 100 feet or so. The vent is 10 feet diameter and plumbs 70 feet.
[3] The water also contains S 0·32 per cent and NH_4 0·28 per cent.
(5) Otukapuarangi, Roturua Geysers, New Zealand (Auckland). The hot springs and geysers were inactive till 1886 when they came alive between the Bay of Plenty and the Upper Tertiary volcanic mountains south of Lake Taupo. The glaciers among the mountains were more extensive in the Pleistocene epoch.

There would thus seem to be strong reasons for believing that heat is derived from highly heated rocks below, but the water may be of meteoric origin. The analyses of waters more directly obtained from recent volcanoes are shown below:—

	(1.)	(2.)	(3.)	(4.)	(5.)
Free HCl	—	43·96	5·50	—	13·04
Free H_2SO_4	—	—	44·17	5·87	2·48
Cl	57·01	14·95	4·96	47·26	44·52
SO_4	3·38	26·96	31·71	9·50	6·40
CO_3	trace	—	—	—	—
Na	16·65	1·32	3·22	24·14	20·75
K	0·93	0·39	—	1·38	3·03
Ca	18·34	2·05	1·21	0·56	0·22
Mg	0·04	0·94	2·31	0·98	1·02
Al and Fe	0·97	9·10	3·18	6·68	6·58
SiO_2	1·78	0·33	2·21	2·37	1·23
	100·00	100·00	100·00	100·00	100·00
Salinity in parts per million of dissolved matter	7,813	18,060	8,296	26,989	60,023

(1) Water from the Boiling Spring, at Savu-Savu, Fiji. This is a mixed chloride and carbonate water and thus strongly suggests a meteoric origin with sea-water.
(2) From the crater of Idjen volcano, Brook Sungi Pait, Java. The large amount of free hydrochloric acid, etc., does not resemble a meteoric water of any normal type.
(3) Hot Spring of Paramo de Ruiz. The analysis has been recalculated. It is an unusual

type of water, and the region contains the active volcano of Tolimia (18,438 feet) which has been the probable origin of the hot spring.

(4) Hot Pool, volcanic crater of Taal, Luzon, Philippines. The sample was taken from the Yellow Lake in the crater. The high percentage of sodium suggests that sea-water has played a part in the eruptions in the Philippines.

(5) Sample from the Green Pool or Lake, crater of Taal, Luzon, Philippines. The variation in the composition as compared with the previous analysis suggests that it is a mixture of sea-water and a deep-seated water.

It is largely outside the scope of this treatise to enter into any detailed description of volcanic action, but attention must be drawn to the chain of volcanoes which are actually active or only recently dormant in various countries. The chain follows the western coasts of South and North America, and around the northern Pacific Ocean to Japan and the Philippines. Another chain crosses the southern Pacific, Tahiti on one side and New Zealand on the other, and continues westward through the islands of Java and Sumatra, up towards Burma through Barren Island. There then begins another chain of volcanoes in western Baluchistan, which continues to the folded ranges of Persia (Iran) and Armenia and loses itself in Italy. There are volcanoes up the " rift valley ", in Africa, and others in the Atlantic, Teneriffe, Azores, Iceland. There are no volcanoes in the Himalayas. The general association of marine areas and folded mountain ranges is obviously, except in the case of the " rift valley ", closely along sea coasts. But there is considerable evidence to show that great pressure, such as orogenic movements, overthrust faulting, and the like, play an important part in most volcanic areas. The association of water geographically is confirmed by the discharge of steam from volcanoes. The chief gas evolved in an eruption is steam. The hot springs areas of Iceland, New Zealand, Alaska, and Italy are clearly associated with the recent volcanic activity, and the immense thermal region of the Yellowstone, Wyoming, is also related to volcanic activity. The heat is thus easily accounted for and the water can usually be ascribed to surface or meteoric waters which have percolated down or been trapped in sediments which have become deeply buried.

Mineral Veins, Pegmatites, and Metamorphic Rocks

There are few now who doubt the influence of water as the agent in the formation of metalliferous veins and related deposits. Mine waters from the Missouri zinc ore region and around the copper area of Butte, Montana, show salinities of about $1 \cdot 00$ and $10 \cdot 0$ per cent respectively, the former carrying nearly $0 \cdot 25$ per cent zinc and the latter nearly $3 \cdot 8$ per cent copper. Another extraordinary water from a mine tunnel at Idaho Springs, Colorado, contained nearly 8 grams per litre of molybdenum oxide (molybdenum blue?), and 25 per cent of the dissolved matter. The water was green in colour and contained a considerable amount of free sulphuric acid. Experimental work has shown that the sulphides of heavy metals can be dissolved in, or decomposed by water alone. It is believed that the gold in the "banket" of the Rand, South Africa, was deposited by percolating water as it travelled through the pore spaces of the conglomeratic material. While zinc and copper are usually of common occurrence in traces or larger amounts, gold is a very rare constituent. Samples of rock, igneous and metamorphic, may show up to 0·025 per cent copper (average), lead up to 0·02 per cent, silver 25 grains to the ton, and gold perhaps an average of 7 grains per ton of the rock. In three rocks—granite, porphyry, and diabase, J. D. Robinson (*Missouri Geol. Surv.*, vol. 7, 1894, p. 479) found averages of 0·004 lead, 0·009 zinc, and 0·006 copper (in percentages). It is recognized that a cold surface water, highly charged with oxygen, acts quite differently as it descends into the rocks, from a hot water, free of oxygen, which ascends from great depths, although both may be in the same circulating system. However, solution will be accomplished by both the descending, highly-oxygenated water, and by the ascending heated and perhaps acid water.

Assuming, therefore, that the metalliferous minerals, ores, and related types, are leached from igneous and metamorphic rocks of deep seated origin, notice must be taken of the fact that 1 cubic kilometre of granite is capable of yielding from 25,000,000 to 30,000,00 tons of water, which at 1,100° would become 160,000,000,000 cubic metres of steam, excluding perhaps 28,000,000,000 cubic metres of other gases. Sedimentary rocks will contain larger amounts of trapped water in their pore spaces and capillary openings. Now it is well known that recrystallization occurs when rocks are deeply buried and both temperature and pressure become high though perhaps insufficient to cause fusion. One of the most astonishing features relating to the coarsely crystalline rocks known as pegmatites is that their crystals, some 3 to 4 feet across and 20 to 30 or more feet in length (such as some beryls and spodumane crystals), must have grown under remarkable conditions since they are clearly not from molten matter. The conditions appear to have been at depths below those of mineral veins but above those at which true igneous rocks, such as dolerite, are believed to occur. If, therefore, sedimentary rocks, such as shales, are deeply buried, they may pass into slates or suffer a greater degree of metamorphism. Some minerals under such conditions change in volume to other minerals of smaller volume but of greater density, but it is a change by recrystallization. Fosterite, Mg_2SiO_4, plus Anorthite, $CaAl_2(SiO_4)_2$, change into Garnet, $CaMg_2Al_2(SiO_4)_3$, and suffer a diminution in volume of no less than 13·0 per cent, 43·9 + 101·1, or 145·0 to 125·8 (molecular volumes).

Alfred Harker (see his book on *Metamorphism—A Study of the Transformations of Rock Masses*, 1932, pp. 8, 15, 18, etc.) has indicated that "solution is in all respects analogous to melting. The solution of a given mineral in a given liquid is a function of temperature and pressure. It is increased or diminished by rise of temperature, according as heat is absorbed or liberated in the act of solution. . . . The presence of some solvent medium pervading the rocks is therefore to be presumed as an essential part of the mechanism of metamorphism of any kind. . . . The kind of solution to which we make appeal is a local and temporary solution. Bodily dissolved, a rock would lose its identity, yielding not a metamorphosed product but a totally new rock. . . . We are then to conceive a rock which suffers metamorphism as being worked over gradually and piecemeal by the very small quantity of solvent present. . . . The principal solvent which officiates in the metamorphism of rocks is doubtless the omnipresent water. . . . That part of the earth's crust which is the theatre of metamorphism is to be conceived therefore as everywhere permeated by a medium consisting of water with other volatile substances. . . . The growing crystal must make a place for itself against a solid resistance, and is to be conceived as forcibly thrusting its way outward from its starting point. . . . ". Harker has stated : " . . . We may infer with confidence that the more active solvents and mineralizers in metamorphism, CO_2 excepted, are of direct magmatic origin. This is also true in the main of the water itself, for any extensive penetration of surface-water into the heated interior crust of the earth is not an admissible hypothesis. For circulation is confined to very moderate depths, and capillarity necessarily ceases at the critical temperature." However, he himself had also written : " The critical temperature of water is 374° C. and this figure will not be much raised by a small admixture of other volatile substances. The critical pressure is for pure water about 200 atmospheres, equivalent to about 2,500 feet of rocks. . . . The solvent power of liquid water falls off rapidly as the critical point is approached, but there seems to be little information concerning the properties of gaseous water above that point."

Radioactive Waters

There was a time when it was thought that the heat emitted by radium might explain the heated condition of the interior of the earth and the cause of volcanic action. It has been surprising that so few thermal spring waters carry radium in solution. The waters from the uranium ore mines of Joachimsthal, in Bohemia, contain radium, and some springs in Portugal are also true radium-bearing waters. In most other cases the radioactivity detected in thermal and other springs is due to Radon (the emanation of radium) which has a half life period of four days and loses all its radioactivity within twenty-eight days. The hot spring at Bath, in England, is radioactive only from the Radon which escapes with the gases that emerge with water. The radon waters bring with them suspended matter which is deposited as mud and it is often found that the muds are radioactive. Some such muds contain radium and are therefore permanently radioactive, although the spring waters are only radioactive from radon and therefore are not permanently radioactive. There is no question of these radioactive waters being of juvenile or magmatic origin. Their chief function has been as an eroding agent and to extract or remove material from underground and bring it to the surface for subsequent disposal.

Experimental work on samples of granite, basalt, dunite, and other types of rock derived from the crustal and inner shells of the earth's Outer Layer, to a depth of about 50 miles perhaps has shown : (i) that the granitic or *Sial* crust, which coincides with the continents or land areas, averages 4×10^{-12} grams of radium per gram of rock ; (ii) that the basaltic or *Sima* layer, which underlies both the continental areas and the oceanic basins, carries 1×10^{-12} grams of radium per gram of rock ; and (iii) that the dunite or ultra-basic layer below the basalt (*Sima*) layer carries the equivalent of $0 \cdot 25 \times 10^{-12}$ grams of radium per gram of rock. The basaltic layer lies from 10 to 20 miles below the granite layer (deeper under the mountain ranges and shallow under worn-down areas such as continental " shields "), and the dunite layer is from 20 to 30 miles below the basalt layer. It would thus seem that the radioactive elements of the earth's crust are more or less concentrated in the continental platforms. It has also been estimated that of the ninety-two elements recognized in the earth's crust the relative percentages are roughly as follows : oxygen 50 per cent, silicon 25 per cent, aluminium 8 per cent, iron 4 per cent, etc., copper $0 \cdot 01$ per cent, zinc $0 \cdot 004$ per cent, lead $0 \cdot 002$ per cent, thorium $0 \cdot 001$ per cent, uranium $0 \cdot 0004$ per cent, tin $0 \cdot 0001$ per cent, silver $0 \cdot 00001$ per cent, gold $0 \cdot 000001$ per cent, and radium much rarer still ($3 \cdot 6$ parts in 10 million of uranium).

CHAPTER VI.—THE DEPOSITION OF SEDIMENTS

Introductory

Although the term *sediment* may be applied to any matter that separates and settles from any liquid, the fluid in this treatise refers essentially to water, and the solid material normally includes all kinds of rock debris as well as some vegetable and animal matter. Thus, while rock fragments from boulders carried by icebergs to the finest muds which settle as the ooze of ocean beds cover the major portion of the sedimentary deposits, largely river-carried sands and silts, there are other sedimentary deposits. These may range from glacial moraines and frost action screes, to deposits of volcanic dust or ash, which on consolidation around the cones of their ejection become tuffs; these are well known by special names, such as the pozzuolana of Pozzuoli, in Italy, the trass of the Eifel district, in Germany, and the santorin of Santorin, or Isle of Thera, in Greece. The term includes all classes of limestones from coral and foraminiferal to the calcareous tufa or travertine composing stalactites and stalagmites. Under the same category come all kinds of deposits around thermal springs and geysers where the deposit is termed siliceous sinter (as distinct from calcareous tufa, although sinter now generally means silica and tufa means calcium carbonate, but this is not a rigid nomenclature). There are also the sulphur and other sublimate deposits around volcanic vents, such as fumaroles and solfatara, when steam is the main gaseous component of both and sulphur is the chief deposit of the latter. In these sediments there must be included coal seams and oil shales, which contain vegetable debris and animal remains and related deposits. Finally, there are the salt or rock salt deposits from sea water and other saline residues on the one hand, and on the other the mineral veins and cementing deposits in the fissures and pore spaces of underground rocks.

The Stratified Sedimentary Deposits

In the estimates made on page 79, and admittedly very approximate, it was computed that 9 cubic miles of sediment in suspension and 2 cubic miles of dissolved matter was carried annually by the rivers from a land area of about 33,000,000 square miles. The remainder, 24,000,000 square miles, of the land, such as most of Australia, the Caspian and Aral basins, the Sahara, etc., do not obviously contribute to the oceanic deposits. This might be questioned from the point of view of glacier erosion in Greenland and the Antarctic

THE DEPOSITION OF SEDIMENTS 105

and the discharge from submarine springs. However, if the total was taken as 12 cubic miles of total solids (suspended silts and dissolved salts) annually, and the existing proportions of land and sea regarded as more or less constant during past geological eras, say, from Cambrian times, then the total removals from the land would be of the order of 6,000,000,000 cubic miles of solids. This, from a land area of 57,000,000 square miles will mean the erosion of over 100 miles height of land from the land and the deposition of sediments (excluding the dissolved matter which presumably remains in solution) of nearly 32·5 miles thickness of silts in the oceans (140,000,000 square miles). If we allow for deposits of salt and limestone the figure might exceed 40 miles thickness. Allowing for some balancing by sinking under loading and uplift from erosion (unloading) of the land, there may be conceded a difference of 60 miles thickness of rock removed by denudation from the continental areas. This refers to an interval of 500,000,000 years, which is the estimate of the duration of the Geological Record of the history of the life of the earth as gauged from fossil remains of plants and animals.

In that valuable treatise, *The Data of Geochemistry*, 1924, F. W. Clarke had sifted the information then available and calculated that the 300,000,000 cubic miles of oceanic waters owed their salinity to 4,800,000 cubic miles of dissolved salts which had been derived from the igneous rocks of the earth's crust. At least he confined his computations to the sodium (Na) percentages, reckoning an average of 1·14 per cent in sea water and 2·90 per cent in the igneous rocks of this element. After allowing for some of the sodium being trapped in the pore spaces and capillary openings of clays and shales of the sedimentary deposits (he allows 35 per cent of the total sodium of the igneous rocks for this), Clarke estimated that about 84,300,000 cubic miles of igneous rocks had been leached. If this quantity of rock had been taken from the land it would have meant an erosion of about 1·45 miles (thick) from the land surface. This is a thickness of 7,656 feet nearly. And allowing for an increase in bulk of the solid igneous rock (of 84,300,000 cubic miles) of 15 to 20 per cent in its conversion to porous sedimentary material (say, to 100,000,000 cubic miles), the sediments would form deposits 0·72 miles or 3,800 feet thick on the ocean floor. The above data were primarily intended to calculate the age of the oceans from the salt (sodium) they contained, and the age arrived at on this basis was roughly 100,000,000 years. This figure is now recognized as far too small, since computations on the basis of radioactive elements have given estimates of 500,000,000 years to the base of the Palæozoic (Cambrian) and 1,500,000,000 to the base of the Azoic (Precambrian), when the earliest sediments appear to have been laid down. Clarke has shown, however, that the chlorine (Cl) component of salt (sodium chloride) makes a great difference in these calculations. Chlorine is present in igneous rocks to the extent of 0·10 or less per cent, as against 2·90 per cent of sodium. Strictly, therefore, the 100,000,000 years should be multiplied by 29 or 30, which will give 3,000,000,000 years as the age of the oceans, 100,000 feet as the thickness of the marine sediments, and 2,444,700,000 cubic miles as the volume of the igneous rocks which have supplied the chlorine and the sedimentary material.

On the calculation made by R. S. Woodward, a 10 mile depth of the earth's crust, allowing for the oceanic depression of 300,000,000 odd cubic miles and the continental elevation, the volume is roughly 1,633,000,000 cubic miles. The proportions, by mass, of the atmosphere : hydrosphere : lithosphere is 0·03 : 6·83 : 93·14 (in percentages). The proportion of the oceans to the land may be taken as 7 to 93 for quick computations. It has been also accepted that the rocks of the continental (and subcrustal region below the land and continental shelves) are in the proportion of 65 per cent granites and granitic types and 35 per cent of basalts and related basic rocks. In such a total an allowance of 5 per cent is allowed for sedimentary strata (4·0 per cent shales, 0·75 per cent sandstones, and 0·25 per cent limestones). Therefore, the volume of the sedimentary rocks on and about the continents amounts to 81,650,000 cubic miles (in terms of the igneous rocks), which is close enough to 84,300,000 cubic miles above mentioned, but the volume is too small a figure. The other figure, 2,444,700,000 cubic miles, is larger than Woodward allowed (for a depth of 10 miles into the crust), but is in rough agreement with a depth of 12·5 miles below sea-level or a true average subcrustal thickness of 10 miles, exclusive of the continental mass both above and below sea-level to an average depth of — 12,500 feet below sea-level. This was the basis of the calculations on page 36 in estimating the water that is held in the rocks of the subcrustal zone (and in which metamorphosed sediments are believed to occur). If the whole bulk has been subjected to erosion then the 5 per cent allowed sedimentary strata is too low or a great part of the ancient sediments have been reconverted back to the igneous types of rocks as discussed on page 102.

Many measurements of the stratified rocks have been carefully made. These show that strata of the same geological age vary in thickness in the same regions and may be thicker or absent in other countries. The kainozoic or tertiary strata in north-eastern India (Assam) are believed to be 45,000 feet thick. The Upper Palæozoic (Permian) coal-bearing strata of the Indian peninsula, known as the Gondwanas (Damudas) average 6,000 feet in the coalfields of Bihar and Bengal. The entire Gondwanas, upper and lower, are not less than 12,000 feet thick where fully developed. The Assam Tertiary strata were laid down in a marine gulf and a large estuary. The Gondwanas were accumulated in wide river valleys and extensive freshwater lakes. In England the Cambrian and other Lower Palæozoic formations of North Wales are marine deposits, and marine conditions appear to have been frequently repeated in succeeding geological epochs over the region of the British Isles. Similar variations have been noted in other parts of Europe as well as in North America. Charles Schukert (for the U.S.A.) has compiled the thicknesses of the strata as follows : Palæozoic era 111,000 feet, Mesozoic era 86,000 feet, and the Kainozoic (Cainozoic or Tertiary plus Quaternary) 61,000 feet, or a total of 258,000 feet for the historical geological record (since the beginning of the Cambrian epoch). He also shows that the strata are composed of 116,000 feet of shales, 91,000 feet of sandstones, and 51,000 feet of limestones. His totals do not show the thicknesses of lava flows and ash (tuff) beds, which are presumably small. Arthur Holmes has given the averages for sedimentary strata as follows : Kainozoic 73,000 feet, Mesozoic 86,000 feet, Upper Palæozoic 90,000 feet, and Lower Palæozoic 95,000 feet, or a total for the fossiliferous formations of 344,000 feet (without tuffs or lavas). The average works out to from 6 to 8 inches deposited in 1,000 years, for the period since early Cambrian times. Holmes also gives a total of 170,000 feet of older strata, which includes all the Precambrian sediments deposited in perhaps a longer period of time, but this estimate, unless it means that conditions were very different in the Azoic and Eozoic eras, suggests that a great deal of sediment has disappeared (metamorphosed) or is not now recognizable as such.

The above thicknesses, averaging from 260,000 to 350,000 feet for the fossiliferous formations, is, of course, very much more than the 100,000 feet of marine strata calculated in a previous paragraph from oceanic areas. In the present case no indication is given of the extent of the deposits, and it is quite certain that such sedimentary deposits do not extend over the ocean floor. It is even doubtful if they can be considered as covering the entire area of the present land surface which is about a third of the earth's surface. In the alternations of land and sea areas, geographically, it is probable that deposition has, in one form or another, been restricted to a relatively small extent of the earth's surface—deltas and shallow seas and inland waters. Nevertheless, the spread of the geological formations, as found on the continents of to-day, shows that at one time or another in past ages most of the lands of to-day—except areas such as the Canadian Shield, the Indian peninsula, and the Siberian region—have been covered by the sea. But the remains of the sediments that are found are not parts of strata that once covered areas comparable with the Atlantic and Pacific of to-day. The facts, so far as they can be safely accepted, suggest that, while the processes of erosion and deposition have remained substantially the same, the areas from which material has been removed and those where it has been deposited have also remained, relatively, of the same extent, and that our estimated rates of erosion and deposition are probably exaggerated. It appears likely that the actual rates are faster than those computed, but that the computations do not allow enough time for periods of no deposition or even of erosion of deposits which are not now seen for measurement.

Fluviatile and Marine Deposition

Both in river valleys and in the sea the deposition of suspended matter primarily depends on the size of the particles, their density, and the speed of the water current. The greater density of sea water and brine cause delays in the deposition of some fine materials, but in some cases a saline water may precipitate fine suspended material more rapidly than would occur in so-called fresh water. It is therefore the carrying velocity of the water current, either river flow or ocean current, that plays the chief part in the transport of rock debris, etc. A torrent charged with suspended mud will roll down large boulders and sweep pebbles and gravel with ease; and so as the current loses velocity the materials will be sorted and settle as separated material—pebbles where the torrent first loses speed, say, in a delta at the head of a lake, coarse sand a little further along, fine sand and silt in the lake or almost stationary water. Thus, in tracing a single stratum of fluviatile material it may be seen to pass from conglomerate to clays. It may even happen that transported vegetable matter, trunks, and branches, etc., may settle still further out and later become coal, so that a coal seam may often be traced through carbonaceous shales to shale. A limestone may take the place of a coal seam and be due to the current carrying much lime in solution and being forced to precipitate this component in a salt or saline water (delta, lagoon, etc.). Studies of the coal seams and associated strata in the coalfields of Great Britain have shown that, while in many instances there is no doubt that forests grew where the coal is now found (in situ), there are other areas where the evidence in support of a true sedimentary origin (drifted material) is equally clear. The splitting up of the Thick or Ten-yard seam of South Staffordshire as it is traced northward and the thickening of the intervening " stone partings " and its change into coarser material, provide proof of its sedimentary origin. The fossil tree-stumps (*Stigmaria*) found in the associated beds in several localities, Sheffield, Manchester, Glasgow, etc., also give support to the theory of in situ occurrences.

Studies in the coalfields of Bohemia had long ago shown (see *Fauna der Gaskohle und der Kalksteine der Permformation Böhmens*, 1883–1901, by Dr. Ant. Fritsch) that the dismal swamps in which those coal seams were formed were inhabited by aquatic animals. And in the Natural History Museum, London, there are numerous specimens of vertebrate remains from the coal measures of Staffordshire and elsewhere in England and Scotland. Still more common are the fresh and brackish water shells *Anthracosia, Anthracomya, Naiadites, Spirobis,* etc., and of marine forms such as *Spirifer, Productus, Nautilus, Aviculopecten,* etc., but these are rarer. The evidence is that, whether the vegetable matter accumulated where the forests grew in dismal swamps or the plant matter was drifted into lakes or wide river valleys and estuaries, there was aquatic life in the waters, and at times sea water also entered the swamps or lakes. All the Indian Gondwana coals were laid down in river valleys and lakes as drifted material. Animal life, whether reptilian or molluscoid, was remarkably rare, since very few remains have been found. Most of the Indian Tertiary coals, also detrital in origin, were deposited in estuaries and lagoons and often contain marine shells. Above the coal seams of the Punjab and Assam, the strata (of the same Tertiary succession—Eocene) are massive Nummulitic limestones (a foraminiferal limestone consisting almost entirely of calcium carbonate extracted from the sea water by these protozoa). The Carboniferous limestone of England consists largely of brachiopod shells (of *Productus, Spirifer, Athyris, Terebratula,* etc., not to speak of crinoids and corals. Coral limestones are known from the Wenlock limestone (Silurian) and the Corallian (Jurassic) to the reefs of the present seas.

It is necessary to say that the estimates for sedimentary deposition are based on observations of river silt accumulations. These deposits are often characteristically the result of floods and in an unconsolidated condition. The beds of silt may have pore space volumes of 50 per cent filled with water. Compression will occur under the weight of subsequent deposition and the water will be reduced, until a shale is formed, perhaps half the thickness of the original layer of mud and with practically no appreciable porosity. That great teacher and experienced geologist, W. W. Watts, had, on the basis of a denudation of 12 inches of the land surface (57,000,000 square miles) in 1,500 years, estimated that the subsequent sedimentation was confined to the seas bordering the continents. He estimated an area of about 100,000 miles of coastline and a width of deposition of 280 miles (a total equal to F. W. Clarke's, 28,000,000 square miles), and calculated the deposition as 24 inches in the smaller area. This works out to 1 inch in 62·5 years. If this is used as a measure then the data given by Holmes—Kainozoic strata 73,000 feet, Mesozoic 86,000 feet, Upper Palæozoic 90,000 feet, and Lower Palæozoic 95,000 feet—represent periods of deposition of 54,750,000 years, 64,500,000 years, 67,500,000 years, and 71,250,000 years respectively for those eras of geological time. This is a total of 258,000,000 years since the beginning of the Palæozoic era. It has been customary to reckon 500,000,000 years since the earliest Cambrian sediments were laid down and to compute 330,000,000 years for the Palæozoic, 110,000,000 for the Mesozoic, and 60,000,000 for the

Kainozoic eras. Thus, while 54,750,000 is fairly close to 60,000,000 for the Kainozoic there is a great difference between 110,000,000 and 64,500,000 for the Mesozoic, and much divergence between 330,000,000 and 138,750,000 for the Palæozoic. A great part of these differences would be eliminated if the compressed thickness of the older and oldest strata was recognized by allowing slower rates of deposition for the beds as now found (altogether the actual rates at the time of deposition may have been the same throughout geological time).

It is to be remembered that clayey strata, shales, are regarded as representing 80 per cent of all the sediments, with 15 per cent for sandstones and 5 per cent for limestones. Since the clays may have up to 50 per cent (by volume) of pore space and the particles as flakes of matter, it is clear that ultimately, with great compression and elimination of the contained moisture, the shale formed may be of half the volume (thickness) of the original deposit. That sand is subject to diminution of volume by dewatering has been proved many times. The dewatering of the sands under Peterborough cathedral, Northamptonshire, England, led to such subsidence that the tower had to be taken down and rebuilt. In New York City there have been places where subsidence was caused by driving tunnels into the sands below and partly draining these porous strata. Even limestones are subject to shrinkage as a result of dewatering. The case of damage to a farm near S. Hetton, Durham, England, first ascribed to subsidence by colliery workings at a depth of 1,200 feet, was subsequently proved to be due to dewatering the Permian (magnesian) limestone at a depth of 450 feet. Had the subsidence been due to the colliery workings the water-bearing limestones would have dewatered downwards and flooded the mine. From the facts disclosed by R. C. S. Walters (see *The Nation's Water Supply*, 1936, p. 148) it would seem there are potentialities for subsidence in the London area between New Barnet and Victoria and so towards Richmond. In this area water at the rate of several millions of gallons daily is pumped out of the chalk, and the lowering of the water level has been considerable, according to observations in 1878 (Lucas), 1911 (L. J. Willis), and in 1927 (Lapworth).

Shoals and Silting

While the change from coarse to fine sediment may be traced in the same stratum as a result of the velocity of a river current having weakened, it is a common observation to see variations of this nature in successive beds in the same sequence of strata. In one exceptional case a coal seam was directly overlain by a gravel bed (see the illustration given in *Mem. Geol. Surv. India.*, vol. lvi, 1930, plate 5, relating to the Jharia coalfield). In this case, while there is obviously a hiatus, the interval in time between the successive deposits was geologically not considerable since the strata are a conformable series. There is a far greater "break", for instance, in the section in Shropshire, where the Upper Llandovery beds (Silurian) overlie Bala volcanic ash beds which in turn overlie Llandeilo flags (Ordovician). In this case a volcanic interval has intervened and the upper strata are also not quite conformable to the lower beds. However, in silting that we can study, say, in tidal rivers and estuaries, etc., the river and marine currents may not only weaken but be reversed by the change of the tide. In rivers like the Hughli the dangerous sands are known to travel slowly downstream, although the shoal itself remains more or less in the same place. This action is caused by the scour of the tail of the sand at ebb tide, when the current is strongest, but as the whole shoal moves the head of the sand appears to have been washed away. However, at the next full or high tide, when the current is slack, the head of the sand is made up by silting from the river. In large rivers, unaffected by the tides, but subject to periods of great flood followed by very little water, a similar action appears to be in progress, and the depth to which the sand in and under the bed of the river is affected depends on the force of the flood. The great railway bridge over the Ganges at Paksey (between Calcutta and Darjeeling) has piers over 100 feet deep into the alluvium, and yet there appears to be movement in the silt to the very foundations of the piers in the present river channel. In the case of another railway bridge over another large river, J. N. D. La Touche tells the following (see p. 85, *The Young Engineer*, 1935) : " The worst scour always takes place at the downstream side of a pier, and is carried back for about half its length. Scour at this point is often very violent, as the following instance which came under my notice will show. The Permanent Way Inspector on a section of a certain railway was fond of fishing, and the fishermen told him that they always got most fish close to one of the piers of a large bridge on his length as there was a deep hole there where the fish found shelter (during floods). He went to the spot and took soundings with a long bamboo ; he found that he could thrust it a long way underneath the pier itself, which fact he reported to his superior officer. When the river went down the hole filled up and so had eluded observation. The sand was dug away, and it was found that the brick masonry had been worn away to about one-third the length of the pier, leaving an opening right through from side to side."

A discovery of the same kind as that just mentioned was made in the Black Canyon of the Colorado river at the site of the Boulder Dam. The following description is by George E. Barbour (see *The Geographical Journal*, vol. lxxvi, No. 6, 1935, " Boulder Dam and its Geographical Setting "). He wrote (p. 501) : " Until the rock floor of the canyon was cleared the condition of the sand and gravel deposit filling the river bed to a depth of 120 feet was a matter of conjecture.

... Trial borings ... had shown a rock bench or terrace projecting from both sides of the gorge beneath the 40 to 50 feet of gravel lying on the channel bottom. When this rock platform was exposed it proved to have such regularity and sharpness of form that it can have only one meaning, namely that the whole mass of sediment is periodically swept over the canyon bottom, scouring the floor right down to rock bed, and then reburying it at slack water under 50 feet of material. This fact received striking proof when a sawed plank was found reposing directly on the bed-rock buried beneath 40 feet of sand and gravel. A second unexpected feature was a narrow tortuous inner canyon cut some 75 feet deeper down the middle of the main rock-bed. The walls of this canyon within a canyon are quite irregular and often coarsely fluted vertically. The flutings are the result of extensive pot-hole erosion by boulders moving under the swirling current action at times when the bench itself is scoured by flood water. As the pot-holes are enlarged the partitions between give way, the holes become confluent, and the winding chasm results. At the lowest point excavated this inner canyon has a width of less than 10 feet." From the above description it is clear that when the Colorado river was in flood it was acting on the solid rock of its bed down to a depth of over 120 feet from the top of the flood water, but that as the current subsided it first filled up the inner and deep canyon and then covered the rock platform, thus giving no idea of the violence of its action in depth, where it could flush with great force more than 115 feet of sand-filled cuttings. Without such proofs few engineers would be inclined to believe that silting follows sand movements down to depths of 50 and 100 feet below normal bed level at each time of high flood.

Not merely is the depth to which movement of sand under the normal bed of a river is possible fully appreciated but the vast quantity of sand moved is even less realized. On the margin of the Damodar river, in the Jharia coalfield in Bihar, India, collieries have been taking sand for stowing. Originally it was believed that as the sand deposit contained about three years' supply fresh sources of supply would have to be looked for later. These sand deposits have continued to yield sand from the same place for over twenty-five years, and it has only been when a high flood has failed that replenishment has not taken place. Each year the entire deficit has been made up by a few days' high flood, and this rarely fails oftener than once in eight to ten years. A single twenty-four hours' high flood has been quite sufficient to refill a depleted sandbank. Irrigation engineers have had similar, but most unfortunate, experiences with regard to the silting of their storage reservoirs. In the case of the reservoir impounded by the Roosevelt Dam, on the Salt river, Arizona, it has been noted that practically all the silt is deposited when the river comes down in flood. When built in 1910 the reservoir had a capacity of 1,367,000 acre-feet.

PLATE XX.—BONNEVILLE DAM, COLUMBIA RIVER, OREGON STATE.

[Facing page 112

PLATE XXI.—THE NORRIS DAM, TENNESSEE.
This storage dam on the Clinch River is one of the upper works of the Tennessee Valley Authority.

PLATE XXII.—THE WHEELER DAM, ALABAMA.
Otherwise known as the Muscle Shoals dam, on the Tennessee River, and one of the chief works of the Tennessee Valley Authority.

In 1914, as a result of flood discharges in 1910 and 1911, 27,000 acre-feet of silt accumulated and again, due to floods, in 1915 and 1916 there was an increase in silt accumulation of 35,000 acre-feet. Tests made on a stream feeding the Juncal reservoir, Montecito, California, showed that small check dams 5 feet high filled with sand and gravel after one night's very heavy rain.

These experiences have led to greater attention being given to prevent or at least reduce the silt that can be carried into reservoirs during periods of heavy rain. The first consideration is that of preventing serious landslips, either by revetting unstable hillsides or by planting suitable trees on the slopes or at the head of the reservoir. In the case of the Macmillan Dam, on the Pecos river, New Mexico, the growth of tamarisk at the upper end of the lake has practically stopped silting. The dam was built in 1894 and by 1913 the silting was so considerable that it was expected the reservoir would silt up by 1935. The capacity in 1894 was 90,000 acre-feet, it was reduced to 74,000 by 1904, and to 62,000 in 1910, and to 45,000 by 1915, but due to the tamarisk arresting the silt the reservoir still had a capacity of 40,000 acre-feet in 1932. In the United States it is believed that the smaller the ratio of mean annual run-off to reservoir capacity the longer will be the life of a reservoir (against silting up). This kind of empirical formula neglects a sudden heavy flood, which, as already indicated, is the chief cause of silting (not merely in the case of reservoirs but in the open alluvial valleys and plains traversed by a river). Much, of course, depends on the catchment, as to whether the rocks are soft and unstable and liable to yield much silt for a given downpour of rain. The above ratio for various American dams (reservoirs) is—Boulder Dam only $0 \cdot 43$, Roosevelt Dam $0 \cdot 54$, Macmillan Dam, $3 \cdot 2$, etc. It may be pointed out that conditions vary greatly. The Colorado river at Yuma, Arizona, has a catchment area of 242,000 square miles and a maximum flood of 210,000 cusecs (cubic feet per second), while the Betwa river, at Pericha, Central India, has a catchment of barely 10,300 square miles, but discharges a flood of 700,000 cusecs.

The problem of silting, whether it concerns the filling up of reservoirs or shoaling in rivers or the formation of sands and banks in the sea, resolves itself into two main questions—the origin of the silt and the removal of the obstruction. In the case of reservoirs it is essential that the planning and construction must include precautions for preventing large volumes of silt-laden water (floods) entering a reservoir. The steps are, of course, to ensure the stability of hillsides by revetting and afforestation, and to have silt traps or check dams (capable of being cleaned) along those rivers which are known to bring down most silt. The design of the dam often includes discharge mechanism whereby the early and heavily silt-laden part of a flood (which will frequently plunge down at the head of a reservoir like the phenomenon known as *Brech* at the mouth of the Rhone, in Lake Constance, and *La Bataillere*, in Lake Geneva) may be passed through the dam carrying away most of the silt. After this action has served the dam is again closed and the later and cleaner flood water refills the reservoir. In the case of silt-laden rivers, such as the Damodar, the silt may be trapped and utilized for sand-stowing in the collieries. This removal of silt would also remove much of the threat from the river (of leaving its channel) in the Burdwan district. And, furthermore, the James and Mary Sands, in the Hughli, below the confluence of the Damodar, might slowly disappear for want of silt. In the case of shoals at sea, such as the Goodwin Sands, off the Kent coast of England between the North and South Foreland capes, it is difficult to make any comment until the source of the sands (supply) and full particulars of its settling are available.

Deposits of Saline Residues

The solid matter that rivers carry in solution varies according to the soils and rocks which lie in the drainage area from which the run-off rain washes out the soluble matter. Generally this dissolved material is carried into the sea. The nature of the contained salts and the quantity carried depend on tributaries which may dilute the main river water or alter it or add to the amount of solids in solution. If the river discharges into the sea, and the sea water mixes with the ocean, a uniform composition of oceanic water results. Analyses of the averages of ocean waters have already been quoted on page 27. The chief salt deposits, such as those of the Magdeburg–Halberstadt region, known as the Stassfurt salts (in Prussian Saxony), are of Permian age, those of England in Cheshire are of Triassic age, the great deposits of Wieliczka, near Cracow, in Poland, are regarded as Pliocene, and the salt forming in the Kara Bughaz, off the Caspian, indicate how all these deposits of salt result from precipitations as sea water becomes concentrated into brine and bitterns. In the United States of America there are salt deposits of various geological ages—the Silurian rock salt of Michigan, the Permian of Texas, the Jurassic of Utah, the Tertiary of Idaho, and the recent deposits of Oklahoma. These occurrences prove that salt-forming conditions have existed in all geological ages since the beginning of the Palæozoic era, and probably previously. The conditions in the Gulf of Kara Bughaz appear to be typical : a hot and arid region in which a shallow gulf is fed from a large volume of sea water as the gulf waters suffer concentration by evaporation. Experimental work by J. Uiglio, more than a century ago, with Mediterranean sea water, gave the following results (recalculated to a volume of 1 litre and weights in grams):—

Sp. Gr. Density.	Volume in Litres.	Fe_2O_3.	$CaCO_3$.	$CaSO_4$ $2H_2O$.	NaCl.	$MgSO_4$.	$MgCl_2$.	NaBr.	KCl.
1·0258	1·000	—	—	—	—	—	—	—	—
1·0500	0·533	0·003	0·0642	—	—	—	—	—	—
1·0836	·316	—	trace	—	—	—	—	—	—
1·1037	·245	—	trace	—	—	—	—	—	—
1·1264	·190	—	0·0530	0·5600	—	—	—	—	—
1·1604	·1445	—	—	·5620	—	—	—	—	—
1·1732	·131	—	—	·1840	—	—	—	—	—
1·2015	·112	—	—	·1600	—	—	—	—	—
1·2138	·095	—	—	·0508	3·2614	0·0040	0·0078	—	—
1·2212	·064	—	—	·1476	9·6500	·0130	·0356	—	—
1·2363	·039	—	—	·0700	7·8960	·0262	·0434	0·0728	—
1·2570	·0302	—	—	·0144	2·6240	·0174	·0150	·0358	—
1·2778	·023	—	—	—	2·2720	·0254	·0240	·0518	—
1·3069	·0162	—	—	—	1·4040	·5382	·0274	·0620	—
Total deposit	.	0·003	0·1172	1·7488	27·1074	·6242	·1532	·2224	—
Salts in bittern	.	—	—	—	2·5885	1·8545	3·1640	·33	0·5339
Sum	.	0·003	0·1172	1·7488	29·6959	2·4787	3·3172	0·5524	0·5340

Salts Laid Down in Concentration of Sea-water.

It is thus evident that the most insoluble components, ferric oxide and calcium carbonate, are precipitated first, followed by calcium sulphate (gypsum) and then by sodium chloride (common salt), with magnesium sulphate and magnesium chloride, and last comes the precipitate of sodium bromide, while the soluble potassium chloride remains in the bittern.

Ocean water contains an average of 3·5 per cent of solid matter in solution, and if the lagoon was not replenished the deposits would be trifling, whereas the salt deposits referred to in the previous paragraph may be upwards of 2,000 feet thick (at Sperenberg, near Berlin, the thickness was more than 1,000 metres). The supply of sea water has obviously been maintained, as in the case in the Gulf of Kara Bughaz, to make up the loss of water by evaporation from the surface of the gulf. In the Stassfurt, salt deposit the section of precipitates from the top downwards was recorded as follows :—

Drift, roughly 25 feet thick. (Top.)
Shales, sandstones and unconsolidated clays of varying thickness.
Younger Rock Salt, thickness variable, sometimes absent.
Anhydrite ($CaSO_4$), 100 to 250 feet thick, seldom missing.
Salt Clay, 16 to 32 feet thick, rarely absent.
Carnallite ($KMgCl_3.6H_2O$) zone, 48 to 120 feet thick with a layer of salt at top sometimes. In places the carnallite is overlaid by a layer of kainite ($MgSO_4.KCl.3H_2O$) and may, in turn, be overlaid by a mixture of salt and sylvite (KCl), called " sylvinite ". At times the mixture may include kieserite ($MgSO_4.H_2O$).
Kieserite zone, followed by polyhalite ($2CaSO_4.MgSO_4.K_2SO_4.2H_2O$).

Salt. Layers, and then the main or older salt (NaCl) beds with anhydrite. The salt is in layers 3 to 4 inches thick with interstratified anhydrite laminæ 0·25 inches thick (probably representing periods of the deposition—annual ?). These strata are from 500 to 3,000 feet thick (average about 2,500 feet and are computed to have been deposited in 10,000 years).

Anhydrite and gypsum ($CaSO_4.2H_2O$) strata at the bottom. A number of other sulphates have been found in the Stassfurt strata with the salt, including glauberite ($CaSO_4.Na_2SO_4$), epsomite ($MgSO_4.7H_2O$), and celestite ($SrSO_4$). Carbonates are relatively rare, but a bituminous (stinkstone) limestone is found at the base. Native sulphur is not rare, nor are iron pyrites. Bromine and iodine are also found in these beds. In the extensive salt deposits of Texas and Louisiana the association of gypsum with the salt (halite : NaCl) is also accompanied by sulphur, sulphurous gases, and petroleum.

In contrast with the salt deposits from sea water and brine springs there are the alkaline carbonates of inland salt and soda lakes, such as those of Great Salt Lake, in Utah, which is the shrunken remnant of Lake Bonneville, and several lakes in California and Nevada which are the remnants of Lake Lahontan. The region is one in which evaporation has played a great part in the shrinking of the great lakes which existed in the Great Basin of the Western United States in the Quaternary period. Analyses of the water of Great Salt Lake and some of its tributary rivers have been given on page 31. Below are given analyses of some of the lakes of the Lahontan basin—Humboldt river, Humboldt lake, and Soda Lake (Ragtown), in Nevada, and some of the lakes of California—Mono Lake, Owens River, and Owens Lake :—

	Lahontan Basin.			Lakes of California.		
	(1.)	(2.)	(3.)	(4.)	(5.)	(6.)
Cl	2·19	31·82	35·38	23·34	9·49	25·40
SO_4	13·92	3·27	10·50	12·86	15·53	9·89
CO_3	39·55	21·57	15·89	23·42	29·84	22·70
Na	13·63	29·97	35·38	37·93	19·83	37·83
K	2·92	6·54	2·13	1·85	w.Na	2·09
Ca	14·28	1·35	—	0·04	8·92	—
Mg	3·62	1·88	0·21	0·10	3·45	—
SiO_2	9·51	3·53	0·25	0·14	12·37	0·20
Al_2O_3	0·38	—	—	trace	—	—
	100·00	100·00	100·00	100·00	100·00	100·00
Salinity in a million	361	929	113,700	51,170	339	118,830

(1) Humboldt river, Nevada (analysis by T. M. Chatard).
(2) Humboldt Lake, Nevada (analysis by O. D. Allen).
(3) The large Soda Lake, from depth of 30·5 metres (Chatard).
(4) Mono Lake, California (analysis by Chatard).
(5) Owens river, at Charles Butte (average analysis).
(6) Owens Lake (analysis by W. B. Hicks). Contains about 1·89 per cent of B_4O_7 and other analyses show NO_3 and PO_4.

Analyses of the waters of the Caspian and Dead Sea have been given on an earlier page (31). Two cases, those of Sambhar Lake, in Rajputana, and Lonar Lake, in the Central Provinces (both in India), the former in a desert region and the latter in Deccan basaltic lava country, show very similar analyses (made by W. A. K. Christie in each case). The Sambhar Salt Lake shows 52·96 per cent Cl, 5·85 per cent SO_4, and 38·86 per cent Na with 2·19 per cent CO_3. The Lonar Lake waters carry 40·76 per cent Cl, 1·48 per cent SO_4, 39·60 per cent Na, and 17·63 per cent CO_3. The Lonar Lake is an alkaline and salt water with a salinity of 8·28 per cent, while Sambhar is a saline or salt lake water having a variable salinity (dilute when filled by flood water and a brine as the waters evaporate). Sambhar is a great salt producing area for northern India. Alkaline waters (lakes) usually indicate volcanic regions (rocks), while saline waters (lakes) are associated with sedimentary rocks. This is largely true of Lonar and Sambhar lakes although there is a strong salt percentage in the former. Much of the " alkali " incrustations of Californian areas is largely sodium carbonate with some salt and sodium nitrate. Trona or Urao ($Na_2CO_3.NaHCO_3.2H_2O$) separates from the waters of Owens Lake as a common incrustation. The presence of borates at Owens Lake was drawn attention to in the analysis already quoted, but borates and nitrates are uncommon deposits. Boric acid and ammonium chloride are volcanic products, and are given off at some of the fumaroles (soffioni) of Tuscany (Italy) with the jets of steam. Boron nitride although a common artificial substance does not occur naturally, and it has been thought that the decomposition is effected by steam (in fumaroles), but this has been disputed. The borax occurrences of Nevada and south California occur in lake or marsh waters, but the chemical explanations are complex. Sodium nitrate is found in association with the borate deposits in the arid region of south California (and in Atacama Desert, of South America). The largest known occurrences of nitrates are found in northern Chili, where the sodium nitrate (" caliche ") often occurs with salt (salares). However, these Chilean nitrates (from the Atacama and Tarapaca deserts) are saline residues, and though consisting largely of $NaNO_3$ (50 per cent), they may contain NaCl (20 per cent) and KNO_3 (7·5 per cent) with insoluble matter (sand) in large amounts (25 per cent). No satisfactory explanation has been given of the origin of these deposits.

Deposits from Hot Springs and Fumaroles

The commonest type of deposit from dripping water or water which flows in a thin layer over irregularities is that seen in limestone caverns in the form of stalactites and stalagmites on the one hand, or embankments which make small pools. In each case the original water has dissolved the limestone which it subsequently deposits as tufa or travertine. Some of the tufaceous banks grow to a large size, both in length and height, and form natural dams which impound considerable bodies of water, such as the clear blue water lakes of Band-i-Amir, in Afghanistan (beyond Bamian, where the Buddhist rock-hewn figures are carved in the cliffs). The tufa deposited from surface seepages or in caves will include those so-called "petrifying springs" which coat any objects (twigs, birds' nests, etc.) over which the water soaks exposed to the air. It is the same kind of action which produces nodules of calcium carbonate in the soil (the familiar *Kunkur* of Indian engineers) or the "hard pan" that may cement the basal layers of a porous stratum. In the case of "hard pan" the cementing material may be also silica or gypsum. The formation of many deposits of gypsum are tufaceous, but in such cases as have been carefully examined it has been found that sulphuric acid has been liberated from a pyritiferous horizon (shale or coal seam), and the acid water has then acted on the limestone, or vice versa, and deposited the gypsum either in massive granular form or as crystals of selenite. In some cases an outcrop of gypsiferous clay has been proved to pass into pyritiferous shale which underlies a bed of limestone.

Below ground after the depth to which meteoric waters penetrate there is the so-called zone of "cementation" in which the mineral waters deposit the material in the pore of sandstones converting soft strata into hard quartzites (with silica), into calcareous sandstones (with calcium carbonate), and ferruginous sandstones (with ferric hydrate). The subsequent history of the water is not easy to follow, but it will contain a new series of dissolved products and, if it sinks deep enough or is joined by water expelled from porous strata under the pressure of overlying beds, it will be warmer and may become highly heated. Such thermal waters appear as hot springs and become difficult to classify if they occur in volcanic areas or have peculiarities of any kind—radioactivity, leave incrustations of some special type, or have gaseous products of an unusual composition. In most instances thermal springs yield deposits of travertine or calcareous tufa, as near the Pulsating Geyser, Mammoth Hot Springs, Yellowstone National Park; sometimes an ochreous ferric oxide deposit appears, as at Bad Nauheim, Hesse, Germany; often the deposit is a siliceous sinter, as that of Steamboat Springs, Nevada, and many geysers—Iceland, New Zealand, and the United States of America. In some cases the deposit contains nearly 40 per cent of manganese oxide (MnO_2), in others there are traces of many metallic sulphides, suggestive of relationships with ore deposits. The subject of mineral and ore deposits is outside the scope of this treatise, but it is evident that heated waters play an essential part in their formation as in the recrystallization of many deeply buried rocks.

THE DEPOSITION OF SEDIMENTS

With regard to metamorphism as applied to deeply buried rocks and the role which a small, perhaps only 2·0 per cent, proportion of combined water plays, in dissolving a little material and permitting it to crystallize, then dissolving another fraction and permitting it also to crystallize, and so working through the entire mass until the whole rock has crystallized, was referred to on page 102. Some of this water may be finally held in minerals like the micas, but some may be given up and appear to originate as juvenile waters. Thus, it is impossible to say in any broad or general way how to distinguish between juvenile or virginal waters and meteoric or vadose waters in thermal springs or even in volcanic eruptions. The following analyses show the gases collected from fumaroles at Vesuvius soon after the eruption of 1906 ; A and B about 300° C. three months after, and C and D at 250° to 280° C. fifteen months after the eruption :—

Gases from Vesuvius (1906).

	A.	B.	C.	D.
HCl	0·78	trace	none	none
CO_2	11·03	6·68	0·80	0·66
CO	none	none	0·15	0·02
H_2	1·24	trace	0·54	0·02
O_2	3·72	6·00	4·59	3·68
N_2, etc.	15·49	24·88	21·23	17·86
H_2O, vapour	67·74	62·44	72·69	77·76
	100·00	100·00	100·00	100·00

These are undried gases, and the percentage of steam is evident. There is a strong meteoric character about these analyses and it is well known that wells and springs shrink at the time of eruptions when situated near volcanoes. And it has been established that lavas like those from Kilauea, Hawaii, yield water and are by no means anhydrous.

Tinstone and Fossil Wood

In concluding this chapter mention must be made of the fact that cassiterite (SnO_2) is seldom found as an original mineral in rocks, but it appears to be leached or dissolved by percolating water, perhaps thermal waters, and then deposited by replacing some other material, perhaps bone or another mineral, and found as " wood tin " or other deposit, such as sinter, or incrustation from a solution in water. Pseudomorphs of cassiterite after feldspar, etc., are known. Percolating waters, working in a similar manner on buried tree trunks, have carried away the hydrocarbon of the wood and replaced it with silica (opaline), calcite, siderite, etc., while leaving the woody texture, thus producing the common fossil wood which is often abundant in some formations.

PART 3
THE UTILIZATION OF WATER

PART 3.—THE UTILIZATION OF WATER

Chapter VII.—General Considerations

Hydrographic Considerations

Man, like other animals, has drawn supplies of water from rivers and lakes and springs since the dawn of life and, as the race spread and developed, he resorted to digging wells, usually in damp stream beds. Subsequently, as irrigation was practised and the need for larger quantities of water arose, the problems of making deeper wells, impounding water, and conducting water have each required attention. Thus, from early times man studied the science of water. However there is little to show that the engineering aspects of water supply were conducted on satisfactory scientific lines until the Egyptian engineers of 3,000 years ago built a dam across the Nile and cut canals for irrigation and navigation in Lower Egypt. And already before the beginning of the Christian era the Romans were constructing arch-supported aqueducts several miles long to bring water from the hills to their cities. The importance of irrigation has been recognized in most countries, but the construction of dams under proper design, where the height exceeds the width in stone and concrete structures, dates back barely half a century. It is now, during the last twenty-five years, that such multiple purpose projects as that of the Tennessee Valley Authority or those of the Bureau of Reclamation in the Colorado Valley, the Columbia Basin, and other schemes are recognized as national benefits to the country.

All water-supply schemes to-day require detailed study. In the case of storage schemes surveys are necessary for the choice of the reservoir site and the selection of the position of the dam. These surveys include the making of topographical maps, elucidating the geological structure of the rocks, and gauging the flow of the rivers and streams. The meteorological conditions of the region have a fundamental effect on a storage scheme since the flow of the rivers depend on rainfall. In such cases as the Nile flood in Egypt the source of the water is the rain on the Abyssinian highlands hundreds of miles away, but it is usually necessary to impound the waters nearer the head of a river than far down its course. It is recognized also that a study of the catchment of a river, particularly in the hills, will discover the chief sources of silt (perhaps from potential landslips) and permit steps being taken to prevent erosion. Where artesian supplies of water are anticipated the geological structure of the basin and the rainfall on the outcropping porous strata are very important considerations if large supplies are to be drawn. In the London basin, due to pumping on the one hand and insufficient replenishment on the other, the underground water-level has fallen appreciably during the past fifty years.

The word hydrographic has normally been applied to marine surveys, such as the making of charts to show depth of water and shoals and dangerous reefs, but it is here employed in a wider sense to include surveys for the storage of water, for canal systems for irrigation or navigation, as well as for hydro-electric projects at natural falls or in connection with dams. The term hydrology, from *hydrol*, the name for the simple molecule of water, might be used, but it is somehow cumbersome in use. The United States Geological Survey publish Water-supply Papers. The Geological Survey of India recognize a Hydrographic Section for attending to water-supply problems. In both cases it is understood that a special branch of study must be established for carrying out the surveying and mapping and the collection of all the necessary data for any water-supply scheme whether for storage, flood control, power generation and navigation from surface supplies, or from wells and borings from underground sources.

Collection of Data

A primary consideration in all problems relating to water is a correct knowledge of the geology of the area that is under consideration. A large-scale geological map with sections to elucidate the structure and a report by an experienced geologist is essential in any inquiry for water, whether for sinking a deep well, driving a tunnel, or founding a dam. These specially named works represent the final selection of a careful survey of a relatively large area. As has been pointed out previously, one well may interfere with another, a tunnel may drain important springs, a dam may be in danger of overturning by faulty siting, a reservoir may quickly silt up by landslips in the catchment area. With the geological considerations must be included the collection of data on rainfall, temperature, winds, and related meteorological information of the region concerned and particularly the area in which the catchment lies. It is important that the meteorological data, the rainfall figures especially, should cover a sufficient period to include years of drought as well as exceptionally heavy rain and floods. Evidence of great floods may often be found during the surveys by the marks on the banks or from local inhabitants. One great flood might wash away the dam and cause devastation more costly than the expense of building the dam. Similarly, if the storage capacity of a reservoir is insufficient to cover a year of drought the famine that might follow may prove disastrous. Thus, without maximum and minimum data the design of the spillways may be too small on the one hand and the capacity of the reservoir inadequate on the other.

Water Supply Factors

Without accurate rainfall and related data it is not possible to ensure satisfactory supplies of water for a given project. In the case of towns and industrial areas it is important that supplies may be augmented as the requirements increase with the growth of the population and expansion of the industries. An adequate supply of good water is the main consideration of any town council or water board. The history of the water supply of Glasgow is of interest, since it was the first of the British reservoir schemes. In 1859 Glasgow drew its water from Loch Katrine by a gravity pipeline 34 miles long on an overall gradient of 10 feet per mile, roughly. A small amount of impounding was necessary and a certain degree of compensation from Loch Vennachar to ensure a supply of 50,000,000 gallons a day. In 1914 the Loch Arklet works were included to improve the storage and to supply 70,000,000 gallons a day to the growing city. Since then other adjustments have been introduced and every effort is being made to avoid embarking on some entirely new project for more water. In the case of Birmingham there is a different story. In 1891 the city drew its supply (15,000,000 gallons a day for 650,000 people) from five streams and six large wells in the Keuper (waterstones) beds. As the supply was neither abundant nor always satisfactory in quality, a scheme was entered upon to draw water from a clean catchment of nearly 120,000 acres in central Wales where the rainfall is about 67 inches annually. Allowing for a local deduction of about 27,000,000 gallons a day (to maintain the Elan river, tributary to the Wye), Birmingham was assured of 75,000,000 gallons a day along a gravity pipeline about 75 miles long, in addition to its original supply.

Previous to 1902 the supply of water to London was provided by several water companies, but from that year the Metropolitan Water Board has been responsible for the water supply of London. The supply in 1933 was estimated at about 294,000,000 gallons a day, of which the Thames provided nearly 200,000,000 gallons a day, the River Lee roughly 45,000,000 gallons a day, and the remainder was chiefly drawn from wells and borings (say 50,000,000 gallons a day). The economic limit of extraction from the Thames has been computed at about 300,000,000 gallons—so as to allow from 170,000,000 to 200,000,000 gallons a day for the flow of the river to be maintained. Since 1931 a supply of 415,000,000 gallons a day has been the objective of the Metropolitan Water Board, and it was nearly achieved by the construction of the 4,000,000,000 gallon capacity reservoir at Staines (which was opened by H.M. the King on the 7th November, 1947). This would allow for about 150,000,000 gallons a day additional to that extracted previously, but it is at the economic limit of the river. As long ago as 1869 a suggestion was made that London should secure pure mountain water from Wales, and schemes were examined in 1894 and 1897 and again, under a Royal Commission, in 1900. It has been estimated that 100,000,000 gallons a day could be obtained at reasonable cost from the Yrfon project in Wales. Any further water for London will probably involve a pipeline scheme from a distance.

The Nile affords a good example of a natural long distance arrangement since this river brings the flood waters of the Blue Nile all the way from the Sudan through a waterless region to Egypt. In the case of the Colorado River Aqueduct, the water is conducted from Havasu lake (impounded by the Parker Dam 155 miles below the Hoover Dam) 242 miles for the supply of 1,000,000,000 gallons a day to Los Angeles and neighbouring cities in southern California. And another 150 miles down the Colorado river, at the Imperial Dam, another take off, the All-American Canal, conveys the water necessary for the irrigation of 400,000 acres 80 miles distant in the Imperial Valley. There are several other irrigation projects in California and Arizona which use or will utilize the Colorado river water between the Parker and Imperial dams. It is to be remembered that the uncontrolled Colorado river, from a catchment of 244,000 square miles and, after a journey of 1,700 miles, discharged 3,000 cubic feet per second of water into the Gulf of California. The discharge at flood was no less than 200,000 cubic feet of water per second. Gaugings had shown capacities of 22,600 cusecs in the Black Canyon (Hoover Dam site), 21,100 cusecs at the junction of the Santa Maria river (Parker Dam site), and nearly 25,000 cusecs at Yuma at the confluence of the Gila river (below the Imperial Dam site). In the last position the flood waters carried immense quantities of silt (estimated at 170,000,000 cubic yards or 320,000,000 tons yearly). There was always danger of the flooded river below Yuma leaving its channel and overflowing into the Salton (sea) depression in southern California (445 square miles in extent and as much as 275 feet below sea-level in places). The control exercised on the Colorado river by the Hoover, Parker, and Imperial dams is thus, basically, to prevent flooding into the Salton sink.

Danger from Contamination

The quality of water is a primary consideration in all supply questions for domestic consumption, whether for farm or village or town. It is a common experience that deep well waters often have an unpleasant taste from the mineral matter contained in solution, even in small amount. A water with a salinity of 15 parts in 100,000 is a " low " mineral water; one with 25 to 100 parts of solid matter in solution in 100,000 parts of water (salinity 25 to 100) is a " moderate " mineral water, and where the salinity ranges from 100 to 200 the water is definitely a mineral water. Sea water averages a salinity of 3,500 parts of dissolved solids in 100,000 parts of water, but the matter in solution is largely common salt. Where the solid in solution is largely lime or calcium carbonate the water becomes " hard "—weak to very hard according to the amount of carbonate dissolved. A " soft " water is relatively pure water, such as might be obtained from a freshwater lake in the Highlands of Scotland or among the mountains of Wales. However, mineral or *inorganic* impurities in solution rarely render a water unsafe or dangerous or really impure for drinking purposes. The danger comes from *organic* matter of sewage, decaying vegetable and animal refuse, which not only impart a disagreeable odour to the water but pollute it with dangerous animal and vegetable organisms and render the water bacteriologically unsafe for human consumption.

A complete examination of a water involves a detailed investigation of its physical condition, its chemical character, and its bacteriological condition. The look of a water is no guide to its purity. A coloured water (peaty matter) may give an unattractive appearance to a safe water. Odour and taste are indications of contamination. Iron compounds and saline matter are easily detected by taste, but such waters may not be contaminated. Marsh waters are always suspect, and certain kinds of bacteria have the property of removing organic matter from waters holding iron compounds in solution. The effect is seen in the precipitation of flocculent red matter, sometimes in sufficient quantity to clog pipes and render the water unusuable. Hard waters on the other hand, may be of considerable value. For example, the water from Rhyaeder, Wales, is so pure as to attack and dissolve lead in the pipes, and to prevent this danger $0·42$ grains of slaked lime are added per gallon of water. This induced " hardness " retards the plumbo-solvency of the pure water. This Welsh water (from the Elan and Clearwen rivers) is often slightly discoloured by peaty matter from the catchment, but this is not the organic contamination which pollutes water. The practice of storing roof rainwater in casks is definitely unhygienic, and sterilization of storage tanks or pipe systems do not prevent water from becoming unhygienic in unclean vessels in a home. The making of wells or the disposal of sewage are matters which require good technical advice. In the former case a seepage of sewage water may be unknowingly tapped. In the latter case the sewage fluids may find their way into water infiltrating into a well. In both cases the consequences of the contamination may be very serious indeed.

Table mineral waters should be colourless, free from sediment and odour. They are often charged with carbon dioxide (as in ordinary soda-water from syphons), but these waters must be filtered and sterilized before adding the carbon dioxide. Freezing is not a sterilizing process and mineral waters must be rendered bacteriologically safe by treatment. Such organisms as *B. typhosus* and *B paratyphosus* may survive as active bacteria in water which has been filtered and frozen, but they are destroyed after two days' heat treatment if the water is warmed to 40° C. (100° F.). These points are only mentioned here, but they are well understood and full particulars will be found in books on water treatment (see the comprehensive treatise on this subject by G. V. James, 1940, which discusses purification of effluents and the treatment of water for all purposes). As previously mentioned, inorganic matter seldom renders water unsafe for drinking, but such substances may be undesirable in different industries. Both " temporary hardness " (due to calcium carbonate) and " permanent hardness " (due to calcium sulphate) should be removed from boiler waters since the dissolved matter is deposited as " scale " in the boiler. Water with some calcium sulphate is an advantage in brewing, whereas magnesium sulphate is an impurity. Pure or " soft " waters are the best and most economical in sugar making, while water with solids in solution cause loss of sugar and may prevent the sugar from crystallizing. Acid waters are always troublesome and should be carefully neutralized with alkaline salts. The greatest care is necessary since these salts tend to accumulate in the water and render it saline.

Stream Gauging and Storage

Many irrigation engineers can estimate very closely the flow of a moderate-sized stream by a careful scrutiny, perhaps assisted by timing the floating of a few twigs over a measured distance where the stream section is easy to calculate. However, while such measurements are useful they are quite insufficient for the computations of important water supply projects. For these larger schemes accurate determinations are necessary, perhaps for two or three years, to secure low and flood water discharges. The data so collected provide information for estimating the make-up of a reservoir from which supplies are to be drawn for town and irrigation and power purposes. In the case of some storage schemes it may take two or more years for the river flow to fill the reservoir impounded by a great dam, such as that of the Hoover (Boulder) Dam in the Black Canyon of the Colorado River. In that case the resulting reservoir, Lake Mead, the largest man-made lake in the world, has a shore line of 550 miles, it has a capacity of 32,359,000 acre-feet (one foot of water on one acre is 43,560 cubic feet), it covers an area of 146,500 acres, its length is 115 miles, and maximum depth 589 feet. The Colorado river is capable of discharging flood water at as high a figure as 240,000 cusecs (cubic feet per second) from the Grand Canyon. The spillways at the Hoover Dam each have a capacity of 200,000 cusecs. Lake Mead has a flood control reserve of 9,500,000 acre-feet. These figures give some idea of the vast quantities that are handled by such a gigantic project. In view of the projects for raising the level of Lake Tana at the headwaters of the Blue Nile in Abyssinia and of a similar scheme for Lake Victoria in Uganda (the Jinja project), the data collected by Captain H. G. Lyons

PLATE XXIII.—THE PARKER DAM, ARIZONA.

This beautiful dam across the Colorado River, between Arizona and California, supplies the aqueduct carrying water to Los Angeles 240 miles away.

[Facing page 128

PLATE XXIV.—DE-SILTING WORKS, IMPERIAL DAM.

These works remove the silt from the water taken by the All American Canal for irrigation purposes in California. The Imperial Dam is on the Colorado River.

PLATE XXV.—STEAM ESCAPING FROM VESUVIUS.

This picture shows the normal discharge of natural steam from a volcano. The " explosions " of most volcanoes are due to a sudden discharge of steam.

[*Facing page 129*

over forty years ago (see *The Physiography of the River Nile and its Basin,* Cairo, 1906) are of the greatest interest at the present time.

From what has been stated and the fact that water supply problems are greater than ever, it is important that methods of stream gauging are understood. The subject is normally treated as a part of hydraulics —dealing with the flow of water in open channels such as canals. The method of stream gauging must depend on the size of the stream, the conditions of flow, and the value of the measurement desired. For medium to small streams the weir is the simplest procedure when it contains a trapezoidal channel for measuring the flow. There are several formulæ for the calculation of the flow, based on the slope of the sides, the nature of the wet surface (bed and sides of the channel), the velocity of flow, and the depth of water in the channel. For very small streams a V-notch device is useful and cheap. For very large streams the construction of a weir may be costly for temporary purposes, and if no simpler means can be devised it is accurate enough to ascertain the cross-section and determine the mean velocity. As explained, the velocity may be gauged by floats (match sticks, twigs, or properly designed objects) timed over a given distance. Coloured chemicals are often used where circumstances allow, but current metres are most commonly utilized. These instruments are unaffected by wind or eddies, and are standardized, and can measure over a considerable time to secure averages at different positions in a river section. They may be used from an anchored boat which can be moved across the river along the line of a cable. In this fashion readings are obtainable at intervals as well as at varying depths of the river section at a given place. As regards the designed " floats ", there are surface floats, sub-surface floats (for required depths), and weighted rod-floats to nearly touch the bed of the stream and yet show at the surface.

Measurement of Silt

The procedure of taking samples of the muddy water of a river and allowing the suspended matter to settle in a long glass measuring tumbler gives a completely misleading idea of the movement of suspended material. If samples are taken from various depths of a flooded water considerable differences are evident, but none of these methods conveys any true picture of the scouring action in progress at the sides or in the bed of a river in high flood. Where the bed is of rock, however irregular its surface, the flood water scours out any loose sand or other debris. If there is a small covering of sand and gravel this material may be swept away during a spate, but replaced by a fresh deposit as the velocity of the current slackens. Indeed, a bar or shoal moves from head to toe downstream so far as the material is concerned although the shoal as such keeps its position in a river channel. This is due to erosion of the toe at high flood and build up at the head as the current slackens and deposits silt. A river bed in alluvial ground may be churned up to a depth of 20 feet during a flood and tend to undermine pier foundations, and yet silt up to its normal position as the flood subsides. The flushing action of flood water in a gorge may continue to a depth of 50 feet, as appears to have occurred in the Black Canyon at the site of the Hoover Dam. The vast amount of silt carried along a river bed during periods of high flood has been appreciated only by experimental desilting. The desilting works at the Imperial Dam on the Colorado river, above Yuma, show that 170,000,000 cubic yards or 320,000,000 tons of deposits are carried by the river annually.

Wells and Springs

It is a common result to find that leakage from a canal bed has soaked into the adjacent alluvium and raised the ground water level to the surface and thus rendered the lands marshy or barren (owing to the salts which are left after evaporation). In the Ganges-Jumna valley (plains), in the United Provinces of India, a special system of pumping has been introduced to lower the ground water level as a preliminary for reclaiming valuable land for agriculture. Where wells draw water from alluvial sands or water-bearing strata, there is always the question of capacity. If the wells are heavily drawn upon the water level falls. The replenishment of the underground water is always from the rainfall which percolates into the ground from porous outcrops. The problem of replenishing underground water supplies is fully recognized, but the practice has not yet become general. There are certain features of this problem which may be drawn attention to. Over-irrigated lands, streams and springs are obvious areas or lines of discharge of water, and wells in and about these positions draw on surplus water, and by doing so utilize run-off water which would otherwise continue downstream. The reverse case is that of conducting water, which would otherwise run-off uselessly, to an area where it may soak into the ground and replenish porous strata which can hold water. These cases of replenishment include areas of over-pumping, but in any case surplus flood waters may thus be conserved underground for the replenishment of well supplies.

Artesian Water

When a boring taps underground water at a certain depth and the water rises up in the bore-hole by hydrostatic head, the water thus found is under pressure or artesian conditions. If the water rises to the top of the boring and overflows the case is a typical artesian or flowing well, but the action involved is the same if the water rises but does not overflow from the well. Borings through the London Clay in the

London area penetrate the water-bearing Thanet Sands and the chalk and encounter water under artesian conditions, but the natural pressure is insufficient to bring the water to the surface. Indeed, due to lack of replenishment of the water in the strata and the continuous pumping, the underground water level has fallen and the artesian conditions exist only in the geological structure of the London Basin and in regard to the water supply.

The name " artesian " is derived from the old province of Artois, Pas de Calais, France, where a water-bearing (porous) stratum occurs between overlying and underlying impervious strata, but as the outcrop (intake) of the porous stratum is elevated the water that enters it adds pressure to the water in this stratum underground. When this stratum was tapped by a well or boring the water was under sufficient " head " to emerge at the surface. Perhaps the best known example of artesian conditions is that of the strata under London, and known as the London artesian basin. Here the chalk (cretaceous) is the porous stratum. It is overlaid by the tertiary or London Clay, and underlaid by the cretaceous Gault Clay. These formations are disposed in a shallow structural basin with its deepest point roughly under Chelsea. Originally borings in London to the chalk tapped the water under pressure which forced it almost to the level of the Thames. Now, unfortunately, the pumping has been so heavy, 3·5 million gallons a day, that the ground water level has fallen more than 200 feet. It would appear that the time has arrived for conveying flood water down to replenish the chalk (underground reservoir or aquifer) or else face the possibility of serious subsidences in the London District.

CHAPTER VIII.—WATER SUPPLY ENGINEERING

Multiple Purpose Schemes

Since the Tennessee Valley Authority have shown how the waters of the Tennessee River and its tributaries may be controlled and utilized by the construction of storage dams, electric power generation, development of irrigation, improvement of navigation, and supply to towns and cities, the consideration of multiple purpose projects is a recognized part of the post-war planning in most countries. The Tennessee River is the largest tributary of the Ohio, which it joins at Paducah, in Kentucky. Above Knoxville the river consists of two main tributaries—the Holston and the French Brand rivers. Other important tributaries are the Clinch from Virginia, joining at Kingston, and the Little Tennessee and Hiwassee from North Carolina and coming in between Knoxville and Chattanooga. The Tennessee River, with the above tributaries, flows south-west through Tennessee into Alabama and curves round in this state and flows into western Tennessee on a northerly course into Kentucky and so to the Ohio on the Illinois border. The drainage basin is 44,000 square miles, and forty years ago the Tennessee valley was a sleepy region with backward townships and a little used waterway. Now there are numerous dams with navigational locks below Knoxville and great hydro-electric power plant at all of them. The Kentucky Dam is about 20 miles above Paducah ; the Pickwick Landing Dam is above Savannah (Tennessee) ; the Wilson or Muscle Shoals Dam is 60 miles upstream in Alabama, and the Wheeler Dam another 12 miles or so higher up ; the Guntersville Dam is a further 70 miles upstream (Alabama), the Chickamauga Dam is a few miles above Chattanooga (Tennessee), and the Watt's Bar Dam 50 miles higher, but 20 miles below Kingston and 70 miles from Knoxville. There is the Norris Dam on the Clinch river tributary, the Cherokee Dam on the Holston river, the Douglas Dam on the French Brand river—all in Tennessee. In addition, there are dams at Fontana on the Little Tennessee and on the Hiwassee rivers—both in North Carolina. It is computed that the total expectation of power from the whole scheme will amount to 2,000,000 kilowatts, and that the river will be rendered navigable for 500 miles, from the Ohio to above Chattanooga.

The Colorado river project is of a somewhat different type to that of the Tennessee. The Colorado basin is in an arid region and reclamation of barren lands, flood control, and power generation have been fundamental considerations. In addition to the great dams—the

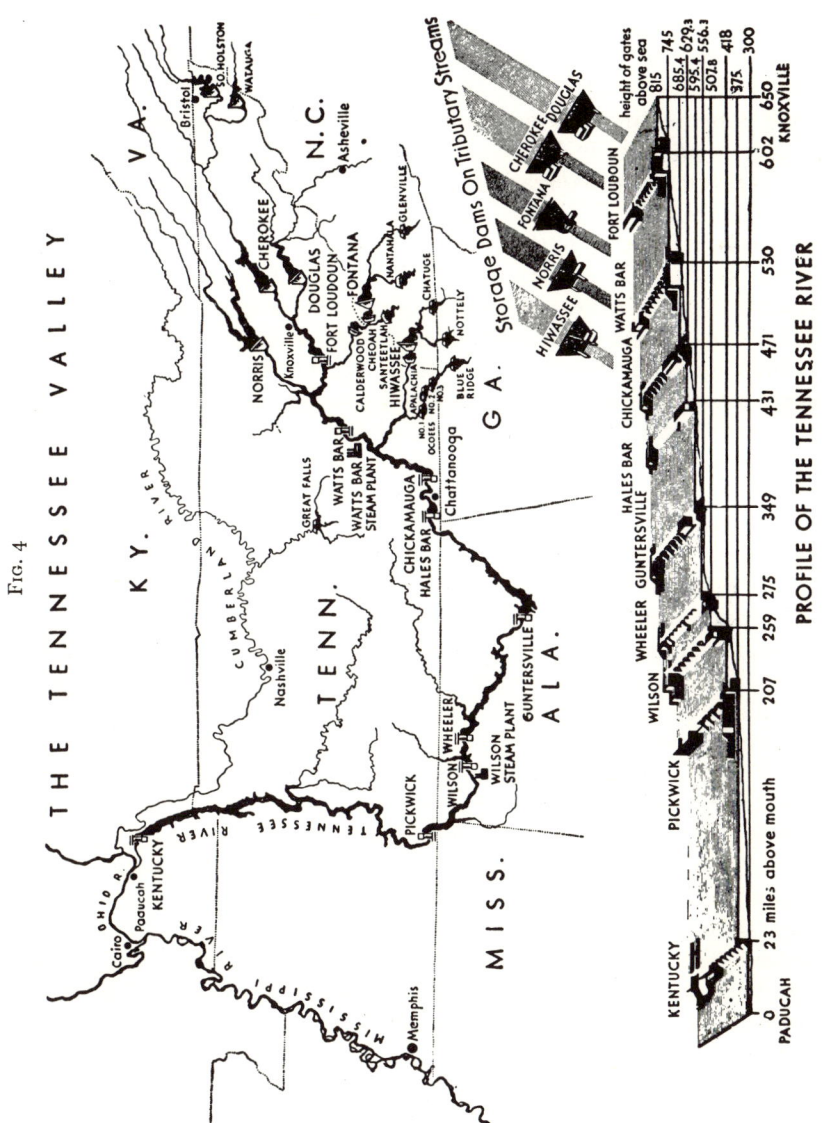

Fig. 4

Hoover, the Parker, and Imperial—and the associated power plant, water supply, and irrigation projects (and new dams—the Davis and Laguna)—the Gila river and its tributaries in Arizona have received attention (by the construction of the Coolidge Dam, on the Gila, and the Roosevelt and Bartlett dams on the Salt river and its tributary). The benefits of the Colorado river reclamation extend from Arizona and California to Nevada, Utah, Wyoming, and Colorado, and the total scheme includes water for many large towns and cities, for irrigation and power, for flood control and desilting, for salinity control and fisheries, for navigation, and also for recreation. The United States Bureau of Reclamation have been engaged upon several other great projects—that of the Central Valley of California, the Columbia River Basin, schemes in Idaho (Twin Springs Dam, on the Boise river), New Mexico (Rio Grande river, Elephant Butte Dam), Wyoming (North Plate river), etc. The multiple purpose dam at Bonneville, Columbia river, 40 miles above Portland, Oregon, has a power plant capable of developing over 500,000 kilowatts. With the completion of the 27 ft. channel from Portland to Bonneville it will be possible for ocean steamers to proceed up river for nearly 200 miles. And as regards the Grand Coulee Dam, the greatest dam so far built, 450 miles above Bonneville, it is to permit of the irrigation of 1,200,000 acres in Washington and generate nearly 2,000,000 kilowatts. It is 550 feet high (three times the height of Niagara), 4,500 feet long, and its spillways are capable of passing a flood discharge of 1,000,000 cusecs. It has impounded waters in a reservoir able to hold 9,610,000 acre-feet. The power will be partly used in lifting water for an irrigation scheme on an elevated area 370 feet above the Columbia river valley.

Whereas the Grand Coulee Dam is of concrete (300,000,000 cubic feet, requiring 2,000 miles of 1 in. piping to cool the setting cement), another dam, Fort Peck (upper Missouri river, in Montana), is of larger bulk, requiring 2,700,000,000 cubic feet of earth (hydraulically placed). It is 242 feet high, 9,000 feet long, and impounds 19,500,000 acre-feet of water. The reservoir floor was dredged to supply the earth for the dam and to increase the capacity of the reservoir. It was constructed for flood control and power generation. The largest low-head power dam was that of the Soviet Government, the Dnieprostroi Dam, across the Dnieper river, U.S.S.R. It was designed to generate 560,000 kilowatts of electric energy when fully developed. Its gates and spillways were to pass a flood discharge of 1,200,000 cusecs. The Assuan (Aswan) Dam, on the Nile, has a height of 175 feet and is 6,400 feet long (built of granite masonry), and impounds 4,000,000 acre-feet of water. The Makwar Dam, near Sennar, on the Blue Nile, impounds 636,000,000 cubic metres of water, while the corresponding Gebel Aulia Dam, on the White Nile, is to impound 4,500,000,000 cubic metres of water. Both are intended to supply water for the Gezira irrigation scheme above Khartoum. One acre-foot corresponds to 43,560 cubic feet; 1 cubic metre is equal to 35·31 cubic feet. The capacity of the Aswan reservoir is thus 174,240,000,000 cubic feet and that of the Gebel Aulia reservoir 158,043,000,000 cubic feet, or very nearly as much as is stored at Aswan. With a dam at the outfall of Lake Tana the supply from the Blue Nile can be considerably augmented, while a dam to raise the height of Victoria Lake must greatly improve the flow from the White Nile.

Power from Rivers

During the past twenty-five years greater and greater attention was given to the utilization of water power, particularly in regions deficient in coal. Switzerland is almost entirely dependent on hydro-electric energy, and shortage of water, either from a deficiency in the rainfall or to severe winter conditions, has an immediate effect on power generation in that country. The simplest mode to generate power is from falls on a perennial river, such as is possible at Niagara Falls, the Victoria Falls on the Zambesi, and in several cases in Norway and Switzerland. In these examples there is no initial outlay for a costly storage dam, which must increase the cost of the power generated. Among the cheapest sources of hydro-electric power are those of Norway, where the estimated production is about 1,000,000 kilowatts and the resources are estimated at 7,500,000 kilowatts. In the United Kingdom, where the conditions are not as attractive as in Norway, nearly £4,000,000 were spent in the Scottish highlands on the Lochaber Water Power Scheme to ensure a steady supply of 75,000 kilowatts to the reduction works of the British Aluminium Company, at Fort William. The project included driving a tunnel 16 miles long under Ben Nevis to carry the water impounded by dams in Loch Laggan and Loch Trieg (see " The Lochaber Water-Power Scheme ", by W. T. Halcrow, *Proc. Inst. C.E.*, vol. 231, 1930–31, p. 54, and also " The Second Stage Development of the Lochaber Water-Power Scheme ", by A. H. Naylor, *Journ. Inst. C.E.*, No. 4, 1936–37).

The water power resources of the world have been estimated at 250,000,000 horse-power (1 horse-power equals 0·746 kilowatts), and of this total the allocation for Canada is 32,000,000 horse-power, at least 55,000,000 horse-power for the United States of America, about the same for South America (54,000,000 horse-power), the British Commonwealth countries (excluding Canada) about 36,000,000 horse-power, with a balance of 73,000,000 horse-power for Europe and the rest of the world. Of this 250,000,000 horse-power (world's reserves), barely 30,000,000 horse-power have been actually developed. In the U.S.A. less than one-fourth of its water power resources have so far been harnessed (and twice as much of the total power generated is obtained from coal-fired steam plant in electric stations). The British Commonwealth (excluding Canada) have probably only developed 1,000,000 horse-power from the available resources. Canada has developed 5,000,000 horse-power out of 32,000,000 horse-power, and South America barely 1,000,000 out of 54,000,000 (horse-power). The proportions in Europe (before the world war of 1939–1945) were as follows : Germany had developed 750,000 horse-power out of 1,500,000 horse-power available from water power ; France 3,000,000 out of 5,000,000 ; Italy 2,750,000 out of 7,000,000 ; Great Britain about 250,000 out of 1,000,000, and Switzerland roughly 1,750,000 out of 4,000,000. In Switzerland there is the remarkable example of a project (Lac Fully) in which a head of 5,412 feet is secured, and the turbines are thus subject to a pressure of nearly 2,405 lb. per square inch. It is not uncommon to operate a hydro-electric power plant in combination with a coal, oil, or gas-fired steam generating plant. This is particularly the case in storage projects in which the water level in the reservoir is liable to fall during a period of dry weather.

Power from Natural Steam

Steam engines have been the most valuable sources of power for industrial development, and 66 per cent of the total power that is generated in the United States is obtained from steam-driven engines. In Italy, where the coal resources are small and hydro-electric power is the chief source of supply, attention was given to the possibilities of natural steam in the hot springs area of Laderello, in Tuscany. As a result of long experience in the extraction of boric acid in the Laderello region, and the use of the steam evolved from the springs, it was possible to erect three steam-driven generators, each of 2,500 kilowatts, in 1916 and to have the plant in successful operation by 1918 (see a paper by Ugo Funaioli, " The Laderello Natural Steam Power Plant," *Engineering*, 10th and 24th May, 1918). At the present time the power plant at Laderello generates 150,000 kilowatts, but further work has been carried out and it is expected to develop 250,000 kilowatts when the additional plant has been installed. These successes in Tuscany have attracted notice in New Zealand, where in North Island the thermal region of Rotorua and Taupo has long been a show place for geysers and hot springs. Investigation has shown that the thermal region of New Zealand is similar to that of Italy (i.e. clearly related to volcanic factors) and that the possibilities for power generation from natural steam under high pressure are attractive. The hydro-electric power plant in the Waikato River project generates a total of 280,000 kilowatts in the region adjacent to the thermal region of Rotorua and Taupo, so that if a similar amount of energy can be secured from natural steam, North Island should become an important industrial part of New Zealand. It is not to be forgotten that similar thermal regions occur in Alaska (Katmai), the U.S.A. Yellowstone Park of Wyoming, and in Iceland. The most famous of the geysers in the Yellowstone National Park is " Old Faithful " (see Photograph VII).

The idea of obtaining power from the internal heat of the earth is an old one. Sir Charles Parsons (of steam turbine fame) put forward a suggestion of this nature, but at a subsequent discussion on the question it was believed that our engineering ability (over thirty years ago) was then incapable of boring to a depth much greater than 8,000 feet. It was also believed that our data on the thermal gradient was insufficient for so experimental an enterprise at that time. It has been reckoned that the average geothermal gradient is about 1° C. per 100 feet descent, but it was already known that in deep mines (Village Deep, on the Rand, and Morro Velho, in Brazil) the gradient was lower 1° C. per 300 feet depth. In thermal regions the gradient is higher than the average, and in the Katmai region, in Alaska, some of the steam discharged from the vents in the valley of Ten Thousand Smokes is definitely super-heated. Furthermore, since mining (with air-conditioning plant) is now conducted at depths of 9,000 to 10,000 feet and borings have been drilled to over 15,000 feet, it would appear that the problem suggested by Sir Charles Parsons can be taken up seriously with every hope of practical success. This does not mean that an area with a low thermal gradient may be selected in preference to a recognized volcanic thermal region. There is no question but that an area where thermal gradient is high, such as the Katmai National Monument, is far more likely to yield more satisfactory results than any other (see Photograph IX).

The deep oil well of the Continental Oil Company, 4 miles west of Wasco, San Joaquin Valley, California, penetrated to a depth of 15,004 feet, at which the strata temperature was found to be 268° F., or 135° C., or well above boiling point under normal atmospheric pressure (see *Bull. Imperial Institute*, vol. xxxvi, part 3, 1938, p. 364). In an article, " Drilling for Steam ; Great Power Potential of Thermal Regions," by A. L. Kidson (*Mining Journal*, 18th December, 1948, p. 945), the following statements are made : " The great danger in boring for steam, of course, lies in the explosive nature of the vapour, accumulated under pressure, and leading sometimes to terrific outbursts when the pressure is suddenly released by the penetration of the drill. To ovrcome this the Italians devised a method of injecting water into the wells, under great pressure, to hold back the steam . . . sometimes these measures proved ineffective, and the wells exploded." As the subject is of considerable interest the following report on " The Recent Eruption of Katmai Volcano, Alaska," by G. C. Martin, might be consulted (see *National Geographical Magazine*, vol. xxiv, February, 1913, No. 2, p. 131), also " The Katmai Region, Alaska, and the Great Eruption of 1912 ", by C. N. Fenner (*Journ. Geology*, vol. xxviii, 1920, p. 569). It was, I believe, Sir Charles Parsons' view that the heat of the strata, not the natural steam, might be utilized. This means that the borings would function as the tubes of a boiler in which water (from the surface) would circulate and become superheated. It is quite a different matter to use the natural steam which might be contaminated with boric and other acids and be liable to explode with the release of pressure made possible by the bore-hole.

Volcanic Energy

In a paper, " Volcanic Energy : An Attempt to Develop its True Origin and Cosmical Relations " (see *Philosophical Transactions*, vol. clxiii, 1873, p. 147), R. Mallet drew attention to the fact that the geothermal gradient is greater in sedimentary rocks and decreases in depth. It was suggested that the heat was due to the pressure of the overlying rocks, and that its diffusion in depth was owing to a smaller percentage of water in the pore spaces, as a result of which the upper strata had a greater specific heat and thus warmer by the heat they retained. The presence of water in the rocks is recognized as a factor in lowering their melting point and so rendering them more fusible. Volcanic explosions, such as that of Krakatoa, in the Sunda Straits, in 1883, are believed to be due to explosions of superheated steam. In case of Krakatoa, an island 5 miles long, 3 miles across, and with a cone 2,700 feet above sea-level, was blown out of the sea. The column of vapour and dust rose to a height of 20 miles, the roar of the explosion was heard at Rodriguez, 3,000 miles away, and walls were cracked and windows broken in Batavia 100 miles distant. The sea waves which spread to the adjacent coasts of Java and Sumatra were over 40 feet high and swept away and drowned upwards of 36,000 people. The energy thus dispayed in a few seconds was enough to scatter 50,000,000,000 tons of rock, much of it to a height of several hundreds of feet.

The Katmai eruption, at the junction of the Aleutian peninsula with the main coast of Alaska, was of a similar character. After a preliminary blowing of incandescent sand over a valley 17 miles long and 4 miles across, an explosion occurred on the 6th June, 1912. It blew away the top of Katmai mountain, leaving a crater 3 miles in diameter and 3,700 feet deep (now filled with water). The mass of rock thus blown away is estimated at 5 cubic miles, or roughly the same amount as that exploded from Krakatoa. In the valley to the north-west millions of vents and fumaroles, from cracks to fissures 120 feet across, appeared, all discharging heated steam and gases. Volcanic dust settled on the island of Kodiak, 100 miles to the southeast, covering the land to a depth of 12 inches. Acid rain fell in Cordova, 360 miles away. Brass was tarnished by the fumes in the air at Victoria, B.C., 1,500 miles to the south. Some of the gases emitted from the fumaroles of this " Valley of Ten Thousand Smokes " as it is now called, were hot enough to melt zinc (420° C.), and even to-day it is possible to fry bacon over some of the vents. The region has been named the Katmai National Monument since 1918. It is a " show place " for tourists. Many theories have been advanced for the Katmai explosion, but the vast amount of steam which has been given up proves that water has been an important factor. Katmai suddenly became a thermal region after the eruption. In the case of the thermal region of the Yellowstone Park, in Wyoming, there appears to be little direct evidence of any volcanic eruptions of recent date, but the plateau is covered by lavas of the same age as those of the tertiary volcanoes, some of which are only recently extinct or only dormant. Mount Lassen, California, erupted in 1914 and has been intermittently active since, but had been dormant for an unknown length of time before. However, the 4,000 hot springs and geysers in the Yellowstone National Park indicate that there is still uncooled lava at no great depth below the surface.

Chapter IX.—Concluding Remarks

The Radioactive Water in Bikini Atoll

A few words of the results of the under-sea atom bomb explosion in Bikini Atoll may be of interest. The results were journalistically described by Chapman Pincher (*Into the Atomic Age*). It was evident by the precautions taken to protect the naval and military scientific observers (who waited five days before venturing to examine the target ships) that the area was unsafe. It was recognized that the waters of Bikini Atoll as well as the target ships were likely to be highly and dangerously radioactive. The contamination of the target vessels was so great that three years after the explosion of the atom (plutonium) bomb under the waters of Bikini Atoll the hulks were still so radioactive as to be unsaleable as scrap iron. As the aerial explosions at Hiroshima, Nagasaki, and at Bikini were different, it has been concluded that the under-water explosion was partly absorbed in producing a radioactive steam which, as a deadly mist, penetrated all the vessels and condensed back into the lagoon and charged its warmed waters also. A submarine explosion of atomic energy would therefore appear to have very serious after effects. Such a bomb exploded in a large reservoir like Lake Mead or Havasu could contaminate all the water supplying the Hoover Dam turbines and generators on the one hand or cause untold misery to the inhabitants of Los Angeles and the cities around it which were supplied through the Colorado river aqueduct. These repercussions might well be of a devastating character if there is no effective means of rendering lakes and reservoir sources of water supply immune to the radioactive contamination experienced at the Bikini Atoll. It would be an easy matter to drop an atomic bomb into a large reservoir, either from a plane or even from a boat, and produce incalculable damage and disaster.

Water Rights

The subject of the ownership of the water rights of a lake, river, or springs is always important, and might easily become a matter for dispute or litigation in a project for water supply. Even so small a stream as the Wandle, which flows from near Croydon and enters the Thames almost unnoticed at Wandsworth, has been the subject of litigation on several occasions. In the case of large rivers and great schemes considerable delays may arise owing to lack of agreement between the parties concerned. Such was the case in the Colorado river project, where agreement had to be satisfactory to several States— Arizona, California, Nevada, and others. In the case of the Grand

Rivers Dam, on the Nescho, Oklahoma, the Governor used the State National Guard to prevent the completion of the dam until satisfactory compensation was arranged for the re-location of roads and bridges which would be necessary when the reservoir should fill. In many cases the compensation takes the form of a portion of the water impounded, as in the case of the Elan river—flow to be maintained from the Rhyaeder reservoir supplying Birmingham with water from Wales. In the case of tidal rivers it is normal for the State to consider the water and the river bed to be Crown property. In other cases the water rights are vested in the parties through whose lands the river flows. Where the stream is itself a boundary then the rights from each bank extend to mid-stream. There appears to be no law regulating the ownership of underground water, since the percolation is not clearly defined in any recognized channel, as is the case on the surface. The problems of riparian proprietors, however, vary with navigable waters, fishery rights, and the extraction of more than reasonable quantities of water along the reach of the owner's property. There must not be any interference with the stream, such as impounding the water by the erection of a weir or dam and thereby causing sensible injury to other proprietors. The right to the waters of a freshwater lake appear to be similar to that for non-tidal streams. Artificial waterways and streams, such as canals and distributaries, come under quite separate treatment.

The comments made in the previous paragraph concern only the practice in England, but there are endless possibilities for misunderstanding and justifiable litigation. Where water is diverted underground to replenish the supply to wells the scope for argument is wide since the problems involved include interference on the one hand and pollution on the other, as extremes. No proprietor can divert a stream to a place outside his property and thus obtain the use of it elsewhere. A similar law appears to have been recognized in the United States. However, the open waters of the Great Lakes are recognized as " high seas ". A State may be qualified to authorize the building of dams for water storage, but if the impounded water covers roads and bridges the flooding becomes a public nuisance and may become a case for intervention by law. The right of acquisition is sometimes exercised in the protection of sites, the construction of works, and the erection of river control structures. Affairs in the Tennessee valley were such that eventually the powers relating to riparian rights were vested in the Tennessee Valley Authority under the central or Federal Government. In the dispute in the Colorado River Project, between Arizona and the six adjoining States, particularly California, Colorado, Nevada, and Utah, special conditions had to be recognized. These include the recognition of a responsible party, properly and publicly declared to be financially and technically qualified, to have established propriety rights. Unfortunately, it seems impossible to avoid qualifying conditions which are liable to produce delays in construction on the one hand or bring about expensive litigation or introduce costs which must render a small project unattractive, or incur both these liabilities, on the other.

The National Aspect of Water Supply

The subject of Water Rights, like that of the International Law on Territorial Waters, is slowly becoming a national one. So far as Scotland is concerned, very interesting opinions have been expressed by the Ministry of Health, Agriculture and Fisheries Department, in a brochure, *A National Water Policy*, issued in 1944. The opening paragraph reads :—

" In this Paper the Government submit to Parliament and the country their proposals for shaping a national water policy. The Paper is concerned with ways and means of ensuring that all reasonable needs for water can in the future be met—and that they can be met speedily and without avoidable waste, either of water itself or of labour, materials, or money." . . . " There is in this country ample water for all needs. The problem is not one of total resources but of organization and distribution. It is an extremely difficult problem, and its complexity is often not appreciated. For the housewife to turn the tap is a simple operation ; its very simplicity belies the immense amount of organization, labour, expense, and ingenuity which lies behind the regular supply of water which she obtains." . . . " Water is a peculiarly difficult commodity to bring to the consumer. It cannot be compressed or concentrated for distribution. It is bulky, and costs of distribution are relatively high. For these reasons local sources must be used as fully as possible. It is essential to have a sound national pattern of supply and distribution. . . ."

The above White Paper on " A National Water Policy " includes an Appendix A on " The Growth and Development of Public Water Supplies ", and another, B, on " The Influence of Geological Factors on Water Supply in Great Britain ", and a third, C, which deals with the subject of " Compensation Water ". In Appendix C the first paragraph states that " Most of the large water undertakers obtain supplies from impounding reservoirs, constructed by building a dam across a stream at a suitable point where the retention of the water will form a lake . . . when water undertakers seek and obtain powers for the purpose, it is the practice in all local Acts authorizing the works to require them to send down the river or stream for the benefit of the riparian interests a daily and usually regular, flow of water of a prescribed amount . . . ". This qualifying condition is reflected in the case of the Birmingham supply from Wales, since the Rhyaeder reservoir provides water for the flow of the Elan river. It is also recognized in the case of the Thames water for London, where a flow of about 180,000,000 gallons a day must be maintained and extractions from the river cannot encroach on this minimum. In the Scottish White Paper a solution to the water compensation was sought from the Central Advisory Committee (Second Report). The conclusions were :—

" (a) The amount of compensation water should be determined on the merits of each particular case.

(b) In assessing the amount, regard should be had to :—
  (i) The character and flow of the stream ;
  (ii) The extent to which it is used for industries, fisheries, etc. ;
  (iii) The probability of future industrial development ;
  (iv) The protection of the rights of riparian and other landowners ;
  (v) The minimum proportion of the flow below which compensation ought not to be fixed in the interests both of public health and riparian owners."

In concluding this treatise on water, the objective of which has been to conduct the reader from the theoretical to the practical aspects of water supply and utilization, it is almost unnecessary to say the question of assured supplies of water is of increasing importance. The outstanding problems of to-day and of the future appear easy to state and most difficult to satisfy, but they may be considered as follows :—

(a) The supply of water to small groups of people at a distance from a main supply and with inadequate or unsatisfactory local supplies.

(b) The supply of additional water periodically to areas which become short of water due to drought or other cause.

(c) The provision of water for replenishing underground supplies which are being overdrawn or receiving less by percolation.

(d) The absolute control of all effluents, sewage, and the like, which might find its way into surface and underground supplies and pollute them.

(e) The control of all drainage of lands and forests, etc., from which floods and erosion directly affect streams and rivers by silting.

(f) The control of underground water, such as leakage from canals which cause swamps or waterlogging, or affect the working of mines by flooding.

(g) Provision of water through properly constructed aqueducts or pipelines for all purposes—power, irrigation, navigation, and town and city supplies and industrial purposes—through constituted authorities, e.g. water boards or councils.

INDEX

Absolute humidity, 10
Absorption, 57, 85
Abyssinia rainfall, 49
Adelsberg cave, 61, 92
Age of Earth, xviii, 3
Air, analyses of, 10, 11, 21
Air conditioning, 22
Air dissolved, 28
Air pressure, 43
Air temperature, 43
Algonkian times, 3
All American Canal, xv, 126
Altitude of Land, xix
Alums, 15
Amur-Kerulen river, 33
Analyses of Lake waters, 31
Analyses of waters, 12, 14, 27, 31, 80, 81, 91, 97, 99, 116, 127
Annual run-off reservoir capacity, 113
Antarctic continent, 35
Aquifers, 6
Area of Earth's surface, xix
Ararat, 3
Arctic Ocean warm water, 29
Aristotle's " elements ", 4
Artificial rain, xxiii, xxiv
Artesian aquifers, 6, 130-1
Artesian springs, 61
Artesian water, 87, 88, 94, 96, 130
Atlantic Ocean, xix
Atomic energy, 16
Atmosphere, 20
Atmospheric moisture, xx, 6, 23, 42
Aswan Dam, 134
Aswan flood, 53
Average sea water, 27, 90, 126
Azoic times, xiv

Backergunge cyclone, 83
Bahrein Islands, 61, 87
Band-i-Amir, 118
Bandy, Professor O. L., xv
Barbour, George E., 111
Bay of Bengal, 45
Bay of Fundy, 83
Beachy Head, 82
Beadnell, H. J. L., 59
Beardmore glacier, 77
Bending of earth's crust, xxii
Ben Nevis, 46, 135
Berezov mammoth, xiv, 3
Biafo glacier, 76
Bikini atoll, 138
Birmingham water, 125
Black canyon, xiv, 111
Boiling point of water, 7
Bonneville basin, 32, 116
Bonneville Dam, xxv, 134
Bore, tidal, 83
Boulder Dam, see Hoover Dam, xxv, 111
Brech (Rhone), 114

Brontman, L., 29
Bridlington (erosion), 82
Budleigh Salterton, 82

Cambrian seas, xxiv
Carlsbad cavern, 95
Caspian sea water, 29, 117
Catchment area, maximum flood, 113
Chemical analyses of water, 13, 14
Cherrapunji rainfall, 48
Chesil Bank, 82
Chilean nitrate (caliche), 117
Chinook in Montana, 43
Christie, W. A. K., 28
Circulation of air, 42
Clark, W. D., xvi
Clarke, F. W., 11, 13, 27
Cloud, 5
Clouds, 46
Cloudbursts, 48
Coal formation, 108
Coast erosion, 81
Collection of data, 124
Colorado River, 70, 78, 79
Colorado river aqueduct, 126
Colorado valley project, 132, 134
Columbia river, 53, 78, 79
Compensation water, 142
Concentration of sea water, 115
Condensing water, xxvi
Cost of electricity, xxiv
Cyclic salt, 28

Dead Sea water, 31, 117
Deep borings, 96, 137
Density of water, 7
Denudation, 69, 78, 79
Deposits from hot springs, 118
Depth of oceans, xix
Depth of silt movement, 112
Deuterium, 19
Dew, 5
Dew in Oman, Arabia, 44
De-silting works, xv
Dhofar springs, 61
Dielectric constant, 9, 16
Disputes over water, 140-1
Dissolved solids, 12, 14, 27, 79
Dolinas and grotto, 61
Dry Air, 43
" Dry Ice ", xxii
Dull days, 47
Dwyka conglomerate, 77

Earliest water, xiv
Earth Pillars, 73
Earth's crust, bending, xxii
Earthquakes, 84
Ebbing and flowing wells, 94
Egyptian irrigation, xxiv
Electrical resistance, 9

Encyclopædia Britannica, xvi, 19, 25
Eozoic era, 3
Equatorial currents, 41-2
Erosion, xxii, 76, 81, 105
Erosion by frost, 76
Erosion by man, 84
Erosion by waves, 81
Euphrates plain, 3
Evaporation, xx, xxi, 41, 42, 44, 45, 51, 64

Facetted boulders, 77
Falmouth fogs, 42
Filling-up reservoirs, 113
Filter beds, 59
Fluviatile deposition, 108
Föhn in the Alps, 43
Forms of water, 4
Fossil wood, 121
Freezing point of water, 7
Freshwater lakes, 32
Freshwater springs, 95
Frigid zone, 26, 34
Frost, 5, 76
Fumaroles, 117
Fusion by drilling, 96

Gebel Aulia Dam, 134
Geohydrology, xvii
Geological Record, 105
Geysers, 98
Gibraltar, 43, 53
Glacial deposits, 77
Glaciers, 34, 46, 76
Glasgow water, 125
Gorges, 73
Gould, Howard, xv
Grand Canyon, 70, 128
Grand Coulee Dam, xxv, 134
Grand Rivers Dam, 141
Great Lakes as High Seas, 141
Great rivers, 33
Great Salt Lake, 31, 33, 116
Greenland, 35
Grotto and *dolinas*, 61
Gulf Stream, 29, 53
Guntersville Dam, xxv

Hail, 5
Hard water, 127
Havasu Lake, 126
Heat from the sun, 40-1
Heavy rain, 48-9, 54
Heavy Water, xiii, 15, 16, 18
Henle, Maurice, xv
Hoar frost, 5
Holderness coast, 82
Homogeneous atmosphere, 6
Hoover Dam, 126, 128
Hot springs, 63, 97, 98-9, 118, 136
Hughli river, 54, 114
Humidity, 22
Hwang Ho river, 55, 80
Hydration, 6, 62
Hydro-electric power, xxiv, xxv
Hydrogeology, xvii
Hydrognosy, xvii
Hydrography, xvii
Hydrolic cycle, xviii, 40

Hydrollography, xvii
Hydrology, xvii
Hygrometer, 45

Ice, 4, 6, 51
Ice Ages, xiv, 40, 52, 77
Icebergs, 76
Ice caps, 6, 35
Imperial Dam, xv, 126
Imperial Valley, xv, 126
Impermeable strata, 86
Infiltration, 57, 86
Interference between wells, 57-8
Itarar beds, 77

Jaisalmer rainfall, 49
James and Mary Sands, 54, 114
Java Man, xiv
Jeffrey's, Harold, 12
Jinja (Nile) project, 128
Joly, John, 28
Juncal reservoir, 113
Juvenile water, 121

Kaolin, 37
Kara Bughaz Gulf, 114
Katmai explosion, 137, 139
Kilometres to miles, 25
Kunkur, 118
Kuro Siwo drift, 53

Lacey, J. M., 57
Laderello steam, 136
La Bataillere, 114
Lakes and rivers, 30
Lake Constance, 114
Lake Eyre, 61, 88
Lake Geneva, 114
Lake Lahontan, 31, 116
Lake Mead, 128
Lake Michigan, 87
Lake Ngami, 56
Lake Tana, 54, 128, 135
Lake Victoria, 128, 135
Land areas, 24, 25
Land breeze, 43
Latent heat of water, 8, 44
Laterite, 37, 89
Life of reservoirs, 113
Lochaber scheme, 135
Loch Katrine, 125
Loch Vennachar, 125
Lonar Lake, 117
London basin, 131
London climate, 47
London water, 125
Lyons, H. G., 128

Macmillan Dam, 113
Makwar (Sennar) Dam, 54, 134
Mammoth, xiv
Mammoth caves, 92, 95
Manchester rain, 47
Marine deposition, 108
Maufe, H. B., 75
Measurement of silt, 129
Mediterranean current, 52
Meinzer, O. E., xvii, 23

INDEX

Merstham tunnel, 60
Metamorphism, 101, 121
Meteoric water, 98, 121
Mill, Hugh Robert, 12
Mineral springs, 98
Mineral veins, 100
Mineral water, 97, 99, 126
Mississippi sediment, xix, 55, 78-9
Mist, 5
Moisture in atmosphere, xx, 10, 42
Mound Springs, 88, 95
Multiple Purpose Schemes, 132
Murray, Sir John, 12
Murree series, 77
Muscle shoals, xxv

National Geographical Society, xv
National Water Policy, 142
Natural steam, 136
Niagara Falls, xxv, 75
New Zealand, 136
Nile drainage area, 50, 53, 55
Nile-Kagara river, 33
Nile mud, 73
Nile silt to Egypt, 73
Nitrogen in water, 29
Noah's Flood, 3
Normal sea water, 27
Norris Dam, xxv

Ob or Obi river, 33
Ocean currents, 42
Oceanic areas, 24, 25-6
Oceanic waters, 27
Oracle at Delphi, 95
Oxygen in water, 29
Ozonosphere, 20

Pacific Ocean, xix
Palæozoic era, 3, 40
Parker Dam, xv, 126
Percolation, xxvi, 57, 85
Percolating water, 37
Permeable strata, 86
Persian Gulf, 44
Petre, J. Foster, xvi
Petrifying springs, 118
Pleistocene Ice Age, xiii, xiv, xxii, 35, 52, 77
Polar Ice, 34
Pool of Siloam, 94
Porosity and pressure, 96
Pore space volume, 59, 94
Postumia cave, 61, 92
Power from natural steam, 130
Power from water, 135
Precipitation of rain, xx, xxi, 46, 51
Purification of water, 127
Purity of water, 127

Queensland artesian basin, 88, 96

Radioactive water, 102, 138
Rainfall, 6, 46
Rainfall disposal, 40, 50, 64
Rainfall on Nile basin, 50
Rain-making schemes, xx
Rainstorm, 48

Rainy days, 47
Ratios for reservoir life, 112
Red Sea current, 53
Regelation in ice, 76
Relative humidity, 10, 45
Removal in solution, 80, 90
Richmond power station, xxv-xxvi
Rift Valley, Africa, 100
Rio Grande scheme, xv
River sediment, 82
River waters, 27, 80-1, 91
Roosevelt Dam, 112
Rotorua geyser, 136

Sahara Desert air, 46
Saline residues, 114
Salinity, 12, 126
Salt Lake City, 32
Salt in the sea, 28
Salt River, Arizona, xv, 112
Salton Sink, 126
Salts in water, 12, 28
Saville, A. H., xvi
Sea water, 24
Sedimentary deposits, 104, 106
Sediment in rivers, 78
Seiche, 83
Sevenoaks tunnel, 60
Setting of cement, 15
Shoals and silting, 110
Sind Desert, 45
Silt and Scour, 72, 111
Silt in suspension, 78
Silt at Yuma, 126
Silting-up of reservoirs, 113
Size of materials, 70
Size of sand grains, 57, 71
Slaking of lime, 15, 16
Sleet, 5
Snow, 5
Snow Line, 29
Soakage and Slips, 72
Sodium in sea water, 105
Sodury tunnel, 60
Soft water, 127
Solids in solution, 79, 90
Solution of rocks, 90
Specific heat of water, 7, 42
Springs, 92, 130
Springs at Carshalton, 60
Staats, W. D., xv
Stassfurt salt, 114
Steam, 4
Steam generators, xxv
Steam from volcanoes, 100
Storm waves, 84
Stratosphere, 20
Stream gauging, 127
Sublimination, 44
Subsidence, 110
Subterranean springs, 61, 92
Sun, heat from, 40
Sunshine hours, 47

Talchir boulder bed, 77
Temperature gradients, xxii, 43, 96
Thermal conductivity, 9
Thermal regions, 137

INDEX

Thermal springs, 95, 118, 137
Thickness of strata, 106, 109
Thunderstorms, 48
Tide, Bay of Fundy, 83
Tillite, 77
Tinstone, 121
Transporting power of water, 70, 72
Travertine (tufa), 118
Trona, 117
Tronstad, Professor Lief, 18
Tropics (Torrid Zone), 26, 27
Tropopause, 20
Troposphere, 20, 43
True Man (*homo sapiens*), xiv
T.V.A. (Tennessee Valley Authority), 132

Underground rivers, 61, 92
Underground water, 59, 85
Under-tow, 83
Unloading (earth's crust), xxii
Urao, 117
U.S. Bureau of Reclamation, 134
Utah Lake, 31

Vadose water, 121
Vaporization of water, 8, 44
Vapour tension, 10
Vesuvius, 121
Victoria Falls, 75
Victoria Lake, 128, 135
Virginal water, xiii, 121
Viscosity of water, 7
Volcanic explosions, 138
Volcanic gases, 121
Volcanoes, 63, 95, 100
Volga river, 33
Volume of water, 23

Wandle river (Wandsworth), 138
Warm water in Arctic Ocean 29
Water disputes, 140–1
Water divining, 86
Water (origin of name), xvii
Water of oceans, lakes, etc., xx
Water, physical properties of, 7, 8
Water in lavas, 38
Water power reserves, 135
Water Rights, 138
Water in rocks, 36, 38
Water in soils, 36
Water vapour, 6, 9, 42
Waterfalls, 73
Waters, R. C. S., 48
Weathering, 6, 62, 88
Webster, Col. M. L., xv
Weight of water, 12
Weir Report, xxiv
Wells, 130
Wells, interference between, 59
Westerly winds, 42
Wheeler Dam, xxv
Wilkinson, D., xvi
Wilson Dam, xxv
Wood Tin, 121
Wool clouds, 46
Woolly mammoth, xiv

Yangtsi Kiang silt, 78, 80
Yellow River of China, 55
Yellowstone National Park, 98, 100
Young, K. K., xv
Yrfon project, Wales, 126
Yukon River silt, 78
Yuma silt from Gila river, 126

Zambesi river, 75

RENEWALS 458-4574
DATE DUE